Second Edition

ANIMAL BEHAVIOR IN LABORATORY AND FIELD

Edited by

Edward O. Price

State University of New York, Syracuse

and

Allen W. Stokes

Utah State University

W. H. Freeman and Company
San Francisco

NOTE: In the United Kingdom, Denmark, and France, a special license is required for the administration of hormone injections to animals, as in Exercises 7, 18, and 25. In Germany, Ireland, and Italy, such a license is required only for experimenters other than graduates or students who have completed at least two years of study in medicine, veterinary medicine, or science.

Printed in the United States of America

International Standard Book Number: 0-7167-0762-4

9 8 7 6 5 4 3 2

PREFACE

A student's training in the science of animal behavior is incomplete without experience in the laboratory. In the behavior laboratory students receive valuable training in research techniques and acquire a practical understanding of those basic concepts and principles of animal behavior treated in texts, lectures, and class discussions.

The typical approaches to the animal behavior laboratory are three: (1) an emphasis on unstructured independent student projects; (2) the scheduling of structured laboratory sessions; (3) a combination of the two. Advantages and disadvantages can be cited for each of these training procedures. Although the individual project allows the student to pursue a particular interest in depth and provides valuable independent research experience, this kind of concentration necessarily restricts the student's empirical training to but a few of the many subject areas normally covered in animal behavior courses. In addition, as class enrollments continue to swell and budgets dwindle, independent research projects are becoming less practical to administer.

In a structured laboratory setting, the student of animal behavior is exposed to a variety of subject areas and research procedures. To the undergraduate, such experience may be more valuable than in-depth pursuit of a single topic. In addition, since the students in the structured laboratory perform the same experiments, they can benefit from sharing ideas and results with classmates. However, the pre-planned structured laboratory may inhibit one's creativity or originality, and might, perhaps, give the student an unrealistic impression of behavior research. Certainly a series of "polished" two- or three-hour laboratory exercises cannot give the student a realistic impression of the types of problems that normally occur in the course of original research, nor of the patience and dedication necessary to cope with them. Thus, the structured laboratory may limit the educational and emotional experiences normally gained from laboratory research.

Because problems and disadvantages are inherent in both the individual project and structured laboratory approaches, many instructors now offer a combination of the two in their teaching laboratories. Thus students are required not only to conduct a series of proven experiments during a scheduled laboratory period, but also to complete an independent project of their own on a particular area of interest. Each individual thereby acquires some acquaintance with a variety of subject areas and research techniques and also experiences the problems and rewards of original research. Those who find the behavior laboratory exciting and challenging will be encouraged to pursue further study in this field, either in subsequent undergraduate courses or, perhaps, in graduate work.

Animal Behavior in Laboratory and Field is aimed at the undergraduate who is embarking on animal behavior research for the first time in either an introductory or advanced course. The principal objective of the manual is to provide ideas for study rather than "cookbook" formulas for specific projects. It is hoped that the student or instructor will use the specific instructions given in each exercise as a *starting point*, and modify the exercise to suit individual requirements and interests. At the end of most exercises, ideas for additional study are suggested to encourage independent investigation by interested students and to emphasize the "open-ended" approach to the behavior laboratory.

Most of the exercises provide instruction for at least one experiment that can be conducted in one or two scheduled laboratory periods. However, a few topics (for example, biological rhythms) cannot be treated adequately unless more time is allotted. Thus observation and collection of data for these exercises must be scheduled whenever free time is available.

The first five exercises are concerned primarily with the techniques employed in the study of animal behavior, and it is hoped that they will be useful to the beginning student. Most of the exercises described in the manual are to be conducted in the laboratory. Although field exercises are just as valuable, they are more difficult to plan and execute, primarily because control over subjects and test conditions is limited. All of the following exercises have been thoroughly tested by their contributors. Further information on each exercise is provided in the Teacher's Manual; if unusual problems arise, you may wish to write to the individual authors.

This manual is intended as a reference text rather than a workbook. For this reason we have kept the number of tables and illustrations to a minimum. We hope that it will prove a useful reference for researchers and others besides those actually teaching or taking a course in animal behavior.

We would like to thank those members of the Education Committee of the Animal Behavior Society (V. J. DeGhett, M. Harless, and F. R. Lockner) who assisted in reviewing the new exercises submitted. Special gratitude is due Ms. Ruth Piatoff, who typed the various drafts of the manuscript. Finally we extend our deep appreciation to those individuals who gave freely of their time to write the exercises included in this publication.

Our only regret is that we could not republish all of the excellent exercises included in the first edition.

January, 1975

Edward O. Price
Syracuse, New York

Allen W. Stokes
Logan, Utah

CONTENTS

Animal Behavior in Laboratory and Field

THE STUDY OF BEHAVIOR

PETER MARLER
The Rockefeller University

Observation and Description of Behavior

GENERAL ASPECTS OF THE BEHAVIORAL METHOD

At the heart of the modern approach to the analysis of behavior in animals is the problem of description. In the physical sciences there is fairly general agreement on the parameters that must be observed and measured for purposes of analysis. A well-trained investigator confronted with an entirely new problem can anticipate fairly accurately the properties he will look for in seeking to relate the new phenomenon to the general body of knowledge. This is of course an overgeneralization, but the point is that a behaviorist is often unable to anticipate in this way. Rather, he has to approach each new animal with an open mind and to discover anew those parameters of behavior which will lead to a more general understanding. If the investigator is already familiar with close relatives of the new subject, then short cuts are possible. But even here previous assumptions can be dangerous and should not be allowed to dominate observations at this stage.

The first task then is simply to observe the behavior in general, in as many contexts as possible. Since we know that the surroundings can distort and modify behavior, the study should be made under conditions which are as natural as possible. Ideally we would follow the animal around on its daily routines in the wild state. In practice we usually have to compromise. A common solution is to cage the animal as comfortably as possible and allow it to behave in as natural a fashion as it will when left without disturbance. The bulk of behavioral data is gathered in this way, and it is supplemented by whatever field observations are possible. Gradually a picture is constructed of the motor patterns which the animal uses in its daily life, the stimuli—both physical and social—to which it seems to be responsive, and the ways in which the behavior changes with a shift in the physiological condition of the animal. Even a partial picture is better than none at all as a prerequisite for the selection of behavioral parameters for, say, a specific experimental purpose.

The collection of such data is the most arduous and demanding aspect of behavioral study. It requires that you spend much time watching the animals, exposing them to your own presence as little as possible. You will at first be confused by the great variety of complex movements, which follow so quickly upon each other that you cannot see what is happening. Then you will begin to see signs of order: the consistent use, by a bird or a mouse, of a certain posture for fighting or threat situations; a particular configuration of the shapes and colors of the fins of a male fish when it is courting a female. By recording data of this kind, however fragmentary at first, you will eventually begin to predict what an animal is going to do in a new situation, and to discern which behavior patterns are rigidly stereotyped, and which are variable. By relating this to variations in the circumstances under which the behavior occurs, you can begin to understand its motivation, and so on.

You will be able to make only a little progress toward achieving a complete picture of any of the behavior patterns which you study. Nevertheless, as you approach a problem, try to keep the goal of a total picture in mind, forming your own impressions rather than simply checking for behavior patterns that may be mentioned on a hand-out sheet.

TERMINOLOGY

The act of describing inevitably requires the naming of things. This means that we should be aware of what the process of naming implies. Certain assumptions, often quite unconscious, underlie the bestowing of a name. These assumptions are partly a help and partly a hindrance to the achievement of the unattainable goal which we nevertheless seek, a perfect description. Consider an example.

There is a variety of cylindrical, hollow glass objects, each with a flat bottom and a cylindrical neck which is commonly blocked by a cylindrical piece of cork. These glass objects we call bottles. Whenever we call something merely a bottle, we assume that for the purposes of discourse it is unnecessary to indicate the individual properties that distinguish it from other bottles. The use of language for description encourages an emphasis upon the properties that certain objects as a class have in common. Once a name has been created, we tend to overlook variations within the category—a tendency which may be an essential part of the process of perception. The way out of this paradox is to qualify the original name, so that we may have, for example, ink bottles, milk bottles, or wine bottles. For describing objects in our daily life such a system works reasonably well. Instead of continuing with our physical description of the different bottles on a strictly empirical basis, we have taken a short cut and defined them in terms of the liquids which they usually hold—usually, but by no means exclusively. We have in fact a functional rather than an empirical description, which is confused by the fact that milk, for example, is sold in some parts of the world in what we would call wine bottles. To resolve this we must go back and describe what we mean by a wine bottle, beyond just, "a bottle that holds wine"!

It is easy to slip into the habit of taking the same short cut in describing behavior. Yet if you do so, you subtly change the description in a manner that can be exceptionally dangerous and misleading in the analysis of behavior.

This source of confusion can be avoided only by describing on a strictly empirical basis. If you see birds regularly using an upright posture while fighting, call it an "upright posture" and not an "upright aggressive posture": you may find that the same bird uses the same posture in other situations. If the name you bestow includes a functional connotation, it may color your subsequent descriptions in a misleading manner. Even if your original judgment is correct, related species may use the same or a similar posture in quite different circumstances. In several recorded cases, a preoccupation with functional situations has led people to overlook such parallels, simply because they were not expecting to see the posture in another situation. This type of error can be avoided by not using descriptive words that are inherently attached to a particular functional situation, such as "submissive," "aggressive," "appeasing," or "sexual." Only in this way can we build up a body of knowledge in which free cross-reference is possible, both within a species and between species.

THE SELECTION OF SPECIES AND BEHAVIORAL PARAMETERS FOR PARTICULAR PURPOSES

The success of a behavioral investigation depends very much on the selection of a subject appropriate to the area of interest. Obviously a study of birds is likely to be a poor introduction to the use of olfaction in social communication. At a more subtle level, the degree of timidity or tameness needs to be considered in relation to any project you have in mind. A knowledge of the animal's natural history thus is essential in making valid initial selections.

The general descriptive process discussed in the previous section is the precursor of the next stage—selection of parameters suitable for quantification. This is a rather crucial step, and the insights gained from descriptive study are invaluable in ensuring that a maximum of biologically meaningful information will result from a minimum of effort. The literature includes an abundance of papers discussing painstaking and exhaustive measures taken of certain aspects of behavior that may well be useful some day but at present are merely studies that introduce no further hypotheses or insights, and prove only that someone has successfully counted something.

For example, in studies of bird orientation, Sauer and Emlen have shown that some birds can orient themselves in the appropriate migration direction with only the stars to guide them. You decide to test a new bird for this ability by placing it in a circular cage in a planetarium. Now the question is, what do you measure? You can put recording perches around the edge so that you know the sector of the cage to which he goes. But the critical factor might be the direction in which he is facing and not the sector in which he stands. Location may be used to show the night orientation in some birds; direction may be used to show it in others. If you were to measure the wrong response and no other, you would find no orientation.

It is best then, not to be too hasty in deciding to make a quantitative study of only one or two measures of a response. First make general observations of as much behavior as you can. If you are going to spend several months measuring something, it is worth a preliminary week or two to make sure that you are measuring the right things before you start.

Many descriptive studies of an "ethological" type include no explicitly quantitative data at all. We must remember, however, that the building up of descriptions is itself a quantitative process, even though the quantification is intuitive and perhaps partly unconscious. But intuitive processes can be trusted only so far.

As soon as the correlations that we seek to establish become more elaborate or more remote from the original observations (e.g., if we hypothesize that a slight shift of balance between attack and escape tendencies in a new situation is shown, or if temporal correlations between several behavioral responses seem to illuminate some problem) quantitative measures must be used (see Exercise 2).

The ideal to aim for is a proper balance between the use of sophisticated techniques of measurement, sampling, and analysis, and the use of your intuitive powers of observation of the animal, its behavior, and the circumstances in which the behavior takes place. In this way we can all contribute to the advancement of animal behavior as a science while ensuring concentration on problems directly relevant to our understanding of the biology and evolution of specific behavioral patterns.

WILLIAM H. CALHOUN
University of Tennessee

2

Quantification of Behavior

An ethologist studying a "new" species needs to know what behaviors to observe. You should therefore begin study of the species by "just watching" the animal to learn its characteristic behaviors (see Exercise 1). Next, you should prepare a list of specific behaviors with a brief, verbal description of each, so that the list defines the animal's behavioral repertoire. This descriptive phase of ethology is only part of the analysis of behavior; you should go further and quantify the behaviors, so that you can both analyze behavioral sequences and conduct inter- and intraspecific comparisons.

This entire process of identification and quantification of behavior is fundamental to all studies of animal behavior, and this exercise will acquaint you with methods of quantification. You will work in a laboratory rather than in the field (although the general methods apply to field studies as well), and will focus your attention on the methodology rather than on the specific data obtained.

In identifying and selecting behaviors for study, you need to satisfy certain requirements. You must be able to describe each behavior objectively, and you should be able to tell someone else what you are observing so that they can reproduce your results. You can readily quantify behaviors that are unitary or discrete by recording their frequency or magnitude. For rodents, defecation is such a behavior; an observer easily notes when a rat defecates, and can count the number of boli. Some behaviors characteristically occur at numerous, brief intervals (for example, mice groom frequently, but for short periods) and to quantify these types of behaviors, you will want to record both frequency and duration. Keep these suggestions in mind as you develop your behavioral check list: in planning your study, you want to anticipate difficulties in measuring behaviors, and to strive to develop a list which is as complete and practical as possible.

You will need only paper and pencil for the exercise, but you can improve your techniques by using electro-mechanical recording devices. (See Hutt and Hutt, 1970, Chapters 4 through 6, for complete discussion of recording methods.) A time-sampling method improves the accuracy of behavioral observation; this method uses a timing device to pace the observer's recording. For example, a tape-recorded signal delivered through an earphone to the observer marks off fifteen-second intervals, and on hearing each signal the observer notes the behaviors presently being displayed. At the end of a given recording period the checks are totalled, yielding a simple behavioral profile.

Multi-pen event recorders allow for continuous recording of an animal's behavior. Each pen of the recorder is wired to a separate switch of the manual keyboard. When a behavior occurs, the appropriate key is depressed to actuate the recording pen, and by holding the key down until the behavior terminates, you record the duration of each behavior. If you cannot monitor all responses, you can add a second observer who records different responses, thereby obtaining a more complete record of a complex behavioral situation. The recordings provide a permanent record of the observations, and you should first analyze them for *frequency* of occurrence of the various responses, in order to generate a behavioral profile. Next, determine relative *durations* of the behaviors, and finally examine the record for *sequencing* of behaviors.

The novice observer may overlook one important part of any scientific endeavor: the reliability, or accuracy, of the methods of recording. Published studies of animal behavior observations are of little value if other scientists cannot reproduce the results, and you will therefore want to ensure that your methods are reliable. During this exercise you will not have time to determine observer reliability, but you should think about ways of doing so. The simplest method to estimate reliability of the observations is for two observers using the same recording methods to observe the same animal at the same time and record the same behaviors. A simple correlational analysis of the results determines the degree of agreement. If agreement is high (+90 or more), you have a reliable observation method.

You are now ready to try these basic steps: selection of behaviors, practicing quantification, and developing behavioral categories. For best results, devote the first part of the period to simple behavior observation, and do your quantification in the second part of the period.

METHODS

Subjects and Materials

To form a basis for comparing behaviors, study at least three species. Hamsters, gerbils, guinea pigs, rats, and cockroaches all are suitable. Use adult forms, preferably all of the same sex; males are preferred because estrous females are likely to behave quite differently from nonestrous ones, thereby introducing marked variability.

For observation, the animals will be placed in an "open field." This is a box constructed of plywood (or similar material) with a floor marked off into a grid of small squares so that movement of the animal can be measured (Fig. 2.1). A hinged lid, either of clear plastic or with a one-way viewing mirror, is helpful, but not mandatory. A lid can prevent an animal from escaping, and an overhead mirror allows you to observe the animal without its seeing you.

When comparing different species, it is best to use special open fields adjusted to the relative sizes of the animals: each field should be large enough to permit marking off the floor into 36 squares whose sides are 2 to $2\frac{1}{2}$ times the length of the body of the adult animal (for a mouse, each square would have sides 10–12 cm long). Paint the interior of the box some neutral color, and paint division lines of a sharply contrasting color on the floor.

To illuminate the field, install lights in the box or use an overhead lamp; most nocturnal animals (e.g., rats and mice) have rod retinas and cannot discriminate between darkness and dim red illumination. Hence, red illumination will permit you to observe a nocturnal animal under simulated nighttime conditions. (See Figure 2.1 for suggested light installation.) For a small field (smaller animals) two $7\frac{1}{2}$-watt bulbs are needed; larger fields (larger animals) require 40-watt bulbs. Use white bulbs when observing diurnal animals.

Procedure

Plan to complete the exercise in two observation periods (preferably two laboratory periods). This will permit a thorough discussion of your results following the first observation period, and time for any outside reading needed to clear up problems that have arisen before planning observations for the second session.

A. Period 1 Form groups of two to four students for each open field and species. (By rotating open-field assignments every 20 to 30 minutes, several

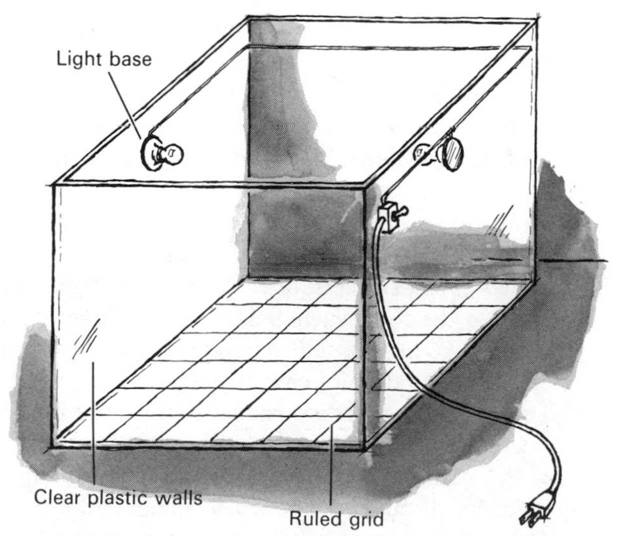

Figure 2-1. Suggested design of open-field observation chamber.

groups can observe different animals and species.) Clean the open fields with a damp paper towel after each animal is removed. A mild vinegar solution (2–4%) eliminates residual odors.

When you are ready to start, place the animal in the field. At first just watch the animal and write down a brief description of each of its behaviors. For example, you might note, "Rat—facial grooming—rubbing face with forepaws alternating with licking forepaws." Avoid using terms that imply causation or function (see Exercise 1). Remember, in this period your concern is not with how frequently a behavior occurs, but with the kinds of behavior.

When all groups have completed their observations, reassemble as a class. For each species, develop a complete list of all the behaviors that were observed. Discuss problems you noted in the course of observing behavior, and suggest solutions. The major aim of the discussion period, however, is to settle on an appropriate and complete list of behaviors for each species so that the frequency and duration of each can be systematically recorded during Period 2. A standard name and description for each behavior should be selected, and a standard observation procedure should be established. Since this exercise is intended to teach you how to develop techniques for observing and recording behavior, there is no "right" result. By the end of this period, you will have had sufficient experience to profit from reading some of the recommended references.

B. Period 2 Using the list of behaviors for each species, prepare a check list for recording your results (Table 2.1). Form into small groups, each group having notebooks, pencils, and a stopwatch. Each group of students may have time to observe only one or two animals, but try to collect data on at least two animals of each species to facilitate species comparisons. Start with dim illumination, making continuous observations for five consecutive 3-minute periods. The grid floor of the open field can be used to quantify ambulation (distance traveled) during the observation period. By comparing the data from the first 3-minute period with that from a later period, it will be possible to determine if the animal's behavioral profile changes with exposure to the apparatus.

After completing the observations and tabulating the results, the groups should reassemble as a class and discuss the similarities and the differences of the species observed and the variability between individuals of the same species. Then review the procedures followed in quantifying behavior and discuss ways you might apply these methods in the field for such a purpose as observing courtship, mating, and nesting in a robin or other tree-nesting bird.

QUESTIONS

1. What are some of the problems encountered in developing a technique for quantifying behavior?
2. What major steps were taken to develop the measures of behavior, and how were the measures recorded? Could you now improve on your methods?
3. Was there any progressive change in behavior as the time an animal spent in the test apparatus increased?

Table 2-1. Check list for recording behavior responses.

Behavior observed	*Frequency or rate of response during successive periods*					*Preceding response*	*Subsequent response*
	1	*2*	*3*	*4*	*5*		
1. Square crossing							
2. Grooming							
3. Sniffing							
4. Rearing							
5.							
6.							
7.							

4. Were there any behaviors that seemed to occur very close together? Could these be lumped into a single class, so that recording only one or two behaviors of this class would imply that the others took place? If so, assign a name to this class.

5. What were the major similarities between species? Differences?

6. Could the methods used in this experiment be employed in the observation of free-living animals?

7. Imagine that you have been asked to obtain a behavioral description of a recently discovered species. Briefly, how would you go about it?

ADDITIONAL STUDIES

If time permits you may wish to observe your animal subjects for additional 3-minute periods. This will facilitate study of the temporal changes in behavior of your test animals in the open field.

You may also wish to observe the influence of light intensity on the behavior of your test species. Test several naive subjects of each species under different levels of illumination. Does increased light intensity inhibit or accelerate the activity of your test subjects?

REFERENCES

Barker, R. G., 1963. *The stream of behavior.* New York: Appleton-Century-Crofts. (See especially Chaps. 1 and 2.)

Bindra, D., 1959. *Motivation: A systematic reinterpretation.* New York: Ronald Press. (See especially Chap. 2.)

Hebb, D. O., 1949. Temperament in chimpanzees. *J. Comp. Physiol. Psychol.* 42:192–206.

Hutt, S. J., and C. Hutt (Eds.), 1970. *Direct observation and measurement of behavior.* Springfield, Ill.: Charles C Thomas.

Schneirla, T. C., 1950. The relationship between observation and experimentation in the field study of behavior. *Ann. N. Y. Acad. Sci.* 51: 1022–1044.

Scott, J. P., 1972. *Animal behavior,* 2d ed. Chicago: University of Chicago Press.

GORDON M. BURGHARDT
University of Tennessee
and Knoxville Zoological Park

Behavioral Study in Zoos and Wildlife Parks

Examining a wide variety of animals is not only interesting but also crucial for an understanding of behavior and its evolution, ontogeny, and ecology. For this reason zoos are becoming increasingly regarded as a windfall of species diversity.

Behavioral study is not the ultimate or major aim of most zoological parks. Therefore, students of behavior must recognize both the advantages and limitations of studying zoo animals, and they must be willing to cooperate with the staff of the zoo at all levels. This exercise, which outlines some strategies specifically designed for zoo or wildlife park studies, is not meant to be a catalogue of the problems to be studied or the methods to be used. Other exercises in this manual, research monographs, and such texts as Hutt and Hutt (1970) serve this function. Many pertinent details and suggestions for zoo research are given by Rabb (in press).

ADVANTAGES AND LIMITATIONS

Besides the traditional zoological garden, aquariums and the newer wildlife parks (Eaton, 1971) offer opportunities for behavioral study. Most of what is said here also applies to working in these environments and hence "zoo" is shorthand for any institution exhibiting captive animals to the public.

A proper zoo, which houses and maintains an array of wildlife, offers diverse opportunities for the study of behavior. Field trips to natural areas to observe wildlife are often time-consuming, expensive, and not at all suited for brief class exercises in which the time actually spent with animals should be maximized. Further, because of season, chance, and daily weather factors it is frequently difficult to observe animals in the wild at fairly close range, if at all. In the zoo some species are always present, behaving and capable of being studied at close range.

Unfortunately, the behavior shown by captive animals is often distantly related to that of their wild relatives. This is much more true of social organization, responses to ecological factors (e.g., climate, prey, predators, etc.), and the frequency, intensity, and duration of behaviors than it is of the basic topography or "structure" of behavior, such as locomotion, grooming, feeding, and fighting. Primates, in particular, evidence this distinction, and much controversy has centered on findings drawn from captive populations. It should be remembered that in general all zoo animals differ from their wild relatives in being: (1) protected from predators, (2) provided with food, (3) limited in exercise, and (4) exposed to a simplified environment.

Notwithstanding these caveats, zoos are exciting places to do research. Many ethologists, psychologists, and anthropologists are turning to zoos and wildlife parks to conduct behavioral observations. Researchers are also attempting to work out methods whereby a species might be better maintained in captivity, or better preserved, or even reestablished in its native habitat. Zoos and similar institutions are the only places where students can see and study exotic species.

THE ZOO TOUR

The zoo tour is a necessary preliminary to a project. Before the tour, make some laboratory observations based on exercises in this manual, and if possible supplement them by film viewing and reading. At the zoo, with the aid of your instructor, become familiar with the variety of species displayed, their taxonomic and ecological relationships, and attempt to describe and analyze some of the behavior patterns observed. Observation of social groups of monkeys, ungulates, wolves, freshwater turtles, and waterfowl can be particularly informative. Discuss with other students and your instructor the behaviors and species observed and the possibilities for further study.

THE PROJECT

Choose a project involving one or more species. Your instructor may have already drawn up a list of possibilities. Projects can be team-coordinated or individual. There are advantages and disadvantages in both.

The team approach allows for more continuous observation by dividing the workload among several persons using the same notation system. For instance, a small group of bears or large cats can be observed for a longer period each day when several students are collecting data. In this way the effect of time of day, feeding, and other factors can be elucidated. Alternatively, a social group of some complexity, such as a "monkey island," can be observed more efficiently. Different students can simultaneously observe different individuals or check the reliability of each other's observations by observing the same animal or phenomenon. Team studies help to focus effort on certain problems, to allow for greater depth, and to facilitate stimulating interchange.

An individual project also has advantages. The student is able to choose a project which depends on his own ability to perform substantial, valid, and replicable observations and experiments. Often, a group effort is dominated by one or more members while other students are restricted from developing their own capabilities and interests. In group efforts scheduling and coordinating observations, data analysis, and the final presentation are often more time-consuming than originally envisioned.

Whether you work alone or in a team should depend upon (1) the species and behaviors being observed, and (2) the interests and personal compatibility of the students. In the comparative studies discussed below, the team approach can be valuable. A good compromise, if enough time is available, is for each student to initiate study on a species, even if others are also observing it. After presenting some initial data and developing his own approaches, he may then want to collaborate and will have a better basis for doing so.

WHAT TO STUDY?

Because the zoo offers a large range of species, behaviors, and problems for study students often do not know where to begin. Frequently, they select species and behaviors for which observation is not feasible, productive, or rewarding. In part this arises from a predetermination to study a behavior (such as courtship) at an inappropriate season, a species that is inadequately housed, or an animal exhibit in which too few or too many interactions normally take place during the time allotted for observation. In deciding on a project, the length of time available (e.g., 2, 4, or 10 weeks) is a crucial consideration, and the instructor will often have to assess the practicality of a given proposal. Be particularly alert to current events at the zoo. Have any new births occurred? Have any of the ungulates begun to mate? Is a pair of swans starting to build a nest? Are new arrivals expected which can be observed adapting to new surroundings? But no matter how well planned a study may be, completely unexpected problems (e.g., disease, inclement weather) can lead to disappointment.

There are many things to be considered when making zoo studies. Where should you observe from? (Don't try to hide—sitting quietly is usually best.) Are binoculars necessary? (Not usually, except in large exhibits, where the public is kept unusually far away, or for details such as facial expressions.) Should notes be written or tape recorded? (Notes and check lists are the most efficient.) Should sunglasses be worn to conceal the observer's eyes, which some animals find threatening? (A good idea with mammals.) Except when two or more individuals are interacting as in courtship or fighting, it is best to focus on one animal at a time.

You will learn much during your project and, if you make another study, you will probably conduct it quite differently. Equipment should be kept to a minimum so that the behavior of the animals is the focus of your study. A pencil, clipboard, and stopwatch are sufficient in most cases (Rumbaugh, 1971).

Listed below are three classes of observational studies that may interest you.

Ethogram. You may want to record and classify the range of behavior patterns which are exhibited by an animal in captivity: the types of social gestures, feeding behavior, locomotion, fighting, mating, spatial organization, etc. Observation of social groups is usually more productive than observation of single or paired animals, although individual recognition is sometimes a problem when observing animal groups. Marking animals is an effective strategy, but it is often not practical for short studies. Usually close observation will uncover certain morphological (e.g., size, scars, deformities, etc.) or behavioral differences which can be used to identify individuals. After a short period of familiarizing yourself with your animal subjects, go to the literature and read about their native habitat, normal patterns of behavior, and the effects of captivity on their behavior. It is always a good policy to read about what other people have done and to see if you agree with their basic classification and descriptions of behavior—but it is much more rewarding to do original work, focusing on one or a few behaviors. If quantitative data on the relative frequency of behaviors for several individuals are desired, be sure that you devote equal time to all animals, not just the more active or "interesting" ones.

Comparative studies. Comparative studies are useful for uncovering ecological and phylogenetic relationships. Comparative studies may utilize either closely or distantly related species. The interpretations of the data obtained will differ depending on the closeness of the taxonomic relationship—i.e., whether they belong to the same genus, or only to the same family, etc.—and on the ecological similarities of their native habitat. You can compare locomotion in various animals, such as the gait in hoofed mammals or climbing in primates. Feeding behaviors can be studied: How do animals with different diets treat their food? What are the relationships between feeding behavior and natural diet? Spatial arrangements are another topic: How do the animals orient themselves in their enclosures? What about territoriality, herding, and habitat preferences? Different species of such groups as felines, apes, birds, and snakes are often maintained in similar habitats, yet close observation will show that the animals do not use the available space in the same way. The influence of the captive environment on observed behaviors should be considered, of course.

Abnormal behavior. In a zoo it is also possible to study animal behavior patterns not normally seen in a natural environment, but which may help the animal adapt to captivity. The most obvious ones are stereotyped movements, such as the pacing of canids or the swinging and climbing paths taken by monkeys. Other examples of abnormal behavior are the begging movements and other responses made by bears and elephants to visitors, keepers, or animals in nearby exhibits (Hediger, 1964). Animals such as ungulates or birds may have formed attachments to humans or individuals of other species, perhaps as a result of imprinting. Many animals mate and give birth in captivity but fail to provide parental care, often engaging in infanticide. Although studies of these topics may not tell us much about the natural behavior of the animals, they may aid in discovering the means of eliminating or controlling the abnormal behavior displayed by animals in captivity.

STUDYING AT THE ZOO

After choosing a project, either you or your instructor should obtain permission to conduct your study at the zoo. Become acquainted with the curator of the general area (e.g., reptiles, ungulates) and the specific keepers who care for the animals to be observed. This has several obvious advantages as well as being simple courtesy.

Remember that the primary function of a zoo is to exhibit animals, in as natural and healthy a state as possible. This means you will not have the freedom to experiment, manipulate, or control the animal to the extent that you would in the laboratory. However, it is often possible to conduct simple experiments. You might be permitted to introduce various objects into the enclosure or present the animals with different types of food, usually more natural or diverse, than are normally fed. Social isolation or the introduction of previously separated animals into the same enclosure, and the rearranging of habitat, particularly with small animals like amphibians, reptiles, and small mammals are often appropriate. Although experiments dealing with the sensory and learning abilities of the animals are also important, they can necessitate rather complex tasks, apparatus, and control procedures to be effective. Essentially, you should begin an experiment only after a period of familiarization and observation. This will prevent disappointment, since it may show you that the experiment cannot be run, and disillusion, since if experiments are begun before gaining enough knowledge about the species and the individual animals you may be unable to do a proper job, in terms of manipulation, experimental design, and interpretation of results.

VISITORS

Visitors to a zoo are often frustrating to students. This is particularly true when people feed the animals, throw objects into the enclosure, strike the glass or bars, move quickly, make abrupt gestures, and yell at the animals and at each other. The crying of children, the odors of hot dogs, popcorn, or perfume can alter the behavior of the animals observed and make interpretations and controlled observations difficult. One possible solution to this problem, of course, is to change the focus from the study of the animals to the study of human reactions to animals or of mutual behavioral patterns. The writings of Hediger (1969) show that there is a source here for much interesting study that can illuminate human behavior.

But if your interest remains with the animals and the disturbances caused by visitors are not violating zoo regulations it may be necessary for you to use the zoo when it is closed to visitors, or at times of the day, particularly early morning, when zoo visitation is minimal. Some buildings, such as reptile houses, are generally closed when the animals are fed, and you may be able to get permission to study the animals then.

If these alternatives are not feasible ask the keeper if it is possible to restrict visitation to the exhibit during your observation or to put up a sign indicating that a project is being conducted and that the animals (including the observer) should not be disturbed. Offer to prepare the sign yourself.

Often when you are photographing or recording behavior, visitors will ask what you are doing. If you are in the middle of a particularly important sequence of behavior the natural tendency is to become annoyed and sharp with the visitor. However, refrain from such impulses and respond cheerfully. Inform the questioner that you are engaged in a study of animal behavior, and that when you have finished what you are doing you will gladly talk about the project. Most observers will thank you and not bother to stay around. Most visitors are at the zoo for a good time and do not want to annoy or upset anyone.

CONCLUDING THE STUDY

Your instructor will undoubtedly require a report of your study. Prepare an extra copy to give to the zoo. The zoo often finds these of interest and importance for their records and for improving exhibits. Indeed, fostering research has become something of a status symbol among zoos. Most zoos welcome suggestions for improving exhibits and a section of your paper should be devoted to the implications of your study for the care and display of the species concerned. In addition, your report will be an aid for students in future classes who may observe the same animals. It will give them the opportunity both to follow up and to extend observations that you began and to have records of specific individual animals stretching over years. This is often lacking in studies of both captive and free-living animals. Such long-term records on individual animals, especially on their ontogeny and social relationships, are becoming increasingly important in ethology.

REFERENCES

Eaton, R. L., 1971. The animal parks: the new and valuable biological resource. *Bioscience.* 21:810–811.

Hediger, H., 1964. *Wild animals in captivity.* New York: Dover.

Hediger, H., 1969. *Man and animal in the zoo.* New York: Delacorte Press.

Hutt, S. J., and C. Hutt, 1970. *Direct observation and measurement of behavior.* Springfield, Ill.: Charles C Thomas.

Rabb, G. B., in press. *Research in zoos and aquariums.* Washington, D.C.: National Academy of Sciences.

Rumbaugh, D. M., 1971. Zoos: valuable adjuncts for the instruction of animal behavior. *Bioscience.* 21:806–809.

Rumbaugh, D. M., 1972. Zoos: valuable adjuncts for instruction and research in primate behavior. *Bioscience.* 22:26–29.

DONALD A. DEWSBURY
University of Florida

Filming Animal Behavior

Motion pictures serve important functions in teaching and research in animal behavior. The purpose of this exercise is to make a film of animal behavior. In so doing, you will learn some basic cinematographic techniques.

Most professional cinematography utilizes 16 mm film. As 16 mm film is expensive, it is not practical for use in a laboratory course. Super-8 or regular 8 mm film—which super-8 is replacing—is less expensive, and involves similar basic cinematographic principles. Purchase and processing of a 50-foot roll of super-8 film costs less than most textbooks and, thus is generally practical for work in a laboratory course.

PREPARATIONS

Logistics of the Project

Behavior films can be made as either individual or class projects. For an individual project, the student may have to provide his own camera. For a class project, the university should provide cameras. With careful scheduling three super-8 cameras and one projector are adequate for small laboratory classes (12 students or under). As super-8 equipment is made primarily for home movies, the cost of such equipment is modest compared to that of much laboratory apparatus.

The author acknowledges gratitude to Dr. Glen McBride who provided the original idea upon which this exercise is based.

The behavior film project greatly expands the range of species that can be in a course and has the further advantage that each student can work with a species and behavioral pattern of his own choosing. Individual film projects and other class projects can be combined even in a ten-week quarter. The final week of the term may be reserved for a "film festival" in which each student presents his film together with background information and narration, followed by discussion. The spirit of the class may be enhanced by setting up an impartial panel to present a modest prize to the student producing the "best" film.

Selection of a Species

The species used is partly a function of the available facilities and equipment. Large diurnal vertebrates are comparatively easy to film. Fortunately, there are many birds, reptiles, amphibians, and mammals within ready access to most campuses. Smaller animals and nocturnal or aquatic species present special problems in filming, particularly in lighting. They might best be left for the experienced cinematographer.

Animals for filming can be found in many settings. Pet dogs, cats, rodents, and snakes can all be good subjects. Many common birds, snakes, and other species, often found in one's backyard, display interesting behavioral patterns. A local duck pond, zoo, or farm may be a source of much interesting animal behavior appropriate for filming. Fish and fowl hatcheries, bird sanctuaries, nature walks, and commercial animal establishments offer great potential. Individuals in departments of anthropology, entomology, psychology, and zoology on your campus probably maintain some interesting species.

Don't overlook humans as subjects. Nonverbal communication, spacing behavior, and other topics in human ethology all are appropriate for this exercise.

Selection of a Behavioral Pattern

In a brief film you will be unable to include all of an animal's behavior and may best concentrate on a limited group of behavioral patterns. Patterns of social behavior are frequently photographed. Agonistic encounters, play behavior, and communication patterns all are of great interest and frequently are readily observable. Many of the most interesting and stereotyped behavioral patterns are to be found in reproductive behavior–including courtship, mating, and parental care. Films of behavioral development and predatory behavior are also quite popular. Don't neglect relatively simple behavioral patterns. Studies of locomotion, spacing, exploration, elimination, ingestion, shelter seeking, and even sleep can be quite effective–particularly when more than one species is included in the film. You may even conduct an "experiment in nature" and record the results on film.

Some successful films in the comparative psychology course at the University of Florida have dealt with nesting behavior in ospreys, a comparison of the strike and ingestion in snakes and lizards, mating in ducks, and copulation in horses. For further ideas, you might review any good textbook on animal behavior or one of the reference books listed at the end of this exercise.

Selection of a Camera

For most projects, the simpler cameras are best. Motor-driven cameras are more useful than spring driven ones, as they are less likely to run down in the midst of important behavioral sequences. A built-in meter system ("electric eye") is helpful to the beginner as it allows him to concentrate on the subject at hand. A zoom lens can be useful in creating special effects and in filming small animals. Cameras that accept film cartridges are preferred over those requiring threading. But however convenient they may be, none of these features is mandatory for the production of a good film.

Selecting Film

Be sure that the film you purchase is appropriate for your camera. For many species, the extra cost of color film may be worthwhile. Films vary in how sensitive or "fast" they are. The new fast films may be worthwhile in situations where you must work in dim light. Be sure that your camera is made to handle them. Slower films work well where lighting is no problem, as outdoors on sunny days.

Preparing Yourself

Shooting the film is best regarded as a late step in film production. Start with library research. It is likely that your library contains some information on the species you select. Careful coverage of the literature may save you much wasted time in the field.

Become familiar with your subjects and their behavior well before you attempt to photograph them. As your film will be very brief (about 3½ minutes for a 50-foot roll), you will have to be very careful about what to include. The object is to have the camera running while the behavior being studied is occurring and to have it turned off at other times. This requires a firm knowledge of the behavior and a keen sense of timing. Some students have found practice with either an empty camera or just a stopwatch to be of value.

Familiarize yourself with your camera before approaching your subjects. Practice focusing. Learn how to obtain proper exposure under a variety of lighting conditions. Be sure that your camera is in good working order and that its lens is clean. Check the batteries and accessories.

Effort expended in observing your animals, practicing with your camera, and in trying to predict their behavior *before* exposing film will usually be repaid many times over in the final product. *Begin early!*

FILMING

Tips on the mechanics of shooting your film can be found in the manual that comes with your camera, or in any of several good books on cinematography. Two good inexpensive booklets from Eastman-Kodak are listed in the reference section below. A few tips of particular relevance will be included here.

Plan the composition of your picture. Try to get close enough to your animals so that they will "fill the frame" as nearly as possible, while being careful that you are not so close as to interfere with the behavior under study. Construction of a simple blind may be useful.

Focusing is critical and has been the most frequent problem encountered with student films. Many cameras require an estimation of distance for focusing. This distance should be estimated as accurately as possible.

Lighting may be a problem if you are filming indoors, particularly if you are using color film. Most indoor lights produce an inaccurate rendering of colors on film intended for outdoor use. See your camera manual for details about appropriate filters, lights, and film if color accuracy is important to you. The behavior of many species is altered in bright light. Thus, you may want to use the minimum acceptable amount of light. As this will require a lens setting that will reduce the depth of field (the range of distance that is in focus), accurate focusing will be of particular importance.

Front lighting of your subjects is almost always preferable for your purposes. If you are filming out-of-doors try to keep the sun at your back. Indoors, top or side lighting may be necessary if glass separates you from your subjects, because front lighting may increase excessive reflections. Be careful to place your lights so that they will not produce heavy shadows on the front of your subjects.

Results are best if the camera is held steady, since jumpy films can be very trying on the viewer. Scenes should not be too short: too many scenes of less than about five seconds can be distracting. Panning (turning the camera across a wide scene) should be avoided if possible; if it is necessary, pan very slowly. Zooming should also be done slowly and the technique should not be overused.

Remember that a standard 50-foot cartridge of super-8 film run at 18 frames per second lasts just three minutes and twenty seconds. Plan your time carefully.

Good film technique is quite important. However, given an acceptable minimum of technique, the success of your film will probably depend on how much of the relevant behavior of your subjects you are able to photograph. Work for the peak behavioral situation by maximizing the likelihood of occurrence of the behavior under study. Pick the right season, time of day, and conditions. After all, it is the behavior that is of primary interest—techniques are valuable only insofar as they help to illustrate behavior.

PRODUCTION

Design your film within the context of some problem or principle. Films of isolated behavior can be interesting, but films which illustrate a principle or phenomenon usually are much more effective.

Have your film processed by a reputable concern and be sure to allow adequate time for processing.

You will want to view your film at least once before showing it in class. If your projector permits, you may decide to show portions to the class at slower than normal speeds. For a class project, editing usually is neither possible nor necessary, although editing can greatly increase the effectiveness of your film. Techniques of editing and splicing are given in most books on motion picture photography, including the two cited in the references to this exercise. Your local photographic dealer may be of help with this and other matters.

What will become of your film when your course is over? If you have a projector and work with super-8 film, you may wish to keep it. Alternatively, you may wish to donate the film to your university. Your instructor may be able to use the film in teaching lecture and laboratory courses in future years.

REFERENCES

Anonymous, 1972. *Better movies in minutes.* Rochester, N.Y.: Eastman Kodak.

Anonymous, 1970. *Home movies made easy.* Rochester, N.Y.: Eastman Kodak.

Bourliere, F., 1964. *The natural history of mammals.* New York: Knopf.

Carthy, J. D., 1965. *The behaviour of arthropods.* San Francisco: W. H. Freeman and Company.

Dewsbury, D. A., and D. A. Rethlingshafer, 1973. *Comparative psychology: a modern survey.* New York: McGraw-Hill.

Ewer, R. F., 1968. *Ethology of mammals.* New York: Plenum Press.

Hafez, E. S. E. (Ed.), 1969. *The behavior of domestic animals.* Baltimore: Williams and Wilkins.

Hall, E. T., 1959. *The silent language.* New York: Doubleday.

Hediger, H., 1968. *The psychology and behaviour of animals in zoos and circuses.* New York: Dover.

Heinroth, O., and K. Heinroth, 1958. *The birds.* Ann Arbor: University of Michigan Press.

Sommer, R., 1969. *Personal space.* Englewood Cliffs, N.J.: Prentice-Hall.

DONALD S. KISIEL
Suffolk County Community College
(New York)

Use of Video-Tape in Analysis of Behavior

The current use of audio-visual media acknowledges the impact that motion pictures have had on the visual-oriented human population. Slow-motion sequences illustrating animal behavior enable the observer to become ever more critical in his analysis and appreciation of various activities. Since teaching and research laboratories have made increasing use of video-tape in the analysis of behavior, students should become aware of the many uses of this technique. Video-taping fills the need for a visual record of behavior as it actually occurred. By repeated analysis of the same behavioral sequence the investigator often makes critical observations not seen the first time through. In addition, repeated data collection on the same behavioral sequence provides the observer with an estimation of data reliability. Interobserver reliability can be tested by comparing data collected by different individuals on the same behavioral sequence.

Filming animal behavior serves the same functions as video-taping (see Exercise 4) and the initial cost is lower, yet there are several advantages of video-taping that should be recognized. (1) The visual record is available for use immediately after taping, because there is no processing delay. (2) The video monitor enables one to view the visual record while the taping is actually in progress, thus enabling mechanical errors to be rectified immediately. Because of the processing delay in cinematography, these same errors may not be determined until it is too late to record the behaviors again. (3) Taping may be done under normal lighting conditions, thereby avoiding bright lights and intense heat from interfering with the subject's behavior. (4) Because video-tape may be reused, it is a less expensive means of recording behavior once the initial equipment has been purchased. (5) Video-tapes are more suitable than film for recording long sequences of behavior since they are capable of longer continuous recording. (6) Video-tape provides a more convenient means of linking vocalizations (auditory signals) with behavioral patterns.

Aside from the initial cost, there are two major disadvantages to video-tape equipment. It is generally less portable than movie cameras, thus confining taping techniques largely to laboratory studies. Unlike movie film, video-tape is not easily sectioned into single-frame visual records to be reproduced as slides and photos, or made into line drawings.

METHODS

Many of the exercises described in this manual are highly suited for demonstrating the uses of video-tape. Your task will be to (1) obtain a 10-minute video-tape record of some behavioral sequence, (2) start an ethogram by identifying (naming) the various behavioral acts observed from the video-tape record, and (3) quantify the frequency and duration (if possible) of one or more selected behavioral patterns.

Your instructor can explain how to operate the video-tape equipment. He may also be able to offer you many hints important for successful taping (e.g., lighting, distance from animals, close focus adjustment, audio and video gain controls, editing). The animal subjects and camera should be situated in a separate room isolated from outside stimulation, especially visual and auditory stimuli (Fig. 5.1). If such a room is not available, you may place the open field and camera behind an opaque screen in a corner of the laboratory near an exhaust fan or some other "masking" noise.

Taping should preferably be done during a scheduled laboratory period but to ensure an "active session" your instructor may have prepared a taped sequence beforehand. (Keep in mind that most ethologists spend days in the field for a few brief moments of activity-filled behavior, and that a captive subject may not always "perform.") If you prefer to pretape a sequence, it may be desirable to tape a variety of animals performing different behaviors in sequence for comparative study.

During taping, use a timing device for temporal measurement. A metronome can be set at 1-second (or $\frac{1}{2}$-second) intervals and the "ticks" recorded on the tape-deck. If a metronome is not available, a verbal "bleep" into the microphone at set intervals will suffice. If the microphone is used during taping to study animal vocalizations, you can time the sequence during a separate replay of the tape and superimpose the timing on the audio portion. Be sure to use the same speed for replay that you used for recording. For more accurate measurements, an electronic digital-readout clock showing elapsed time in seconds (or tenths of seconds) can be placed within the field of view of the studio camera. Duration of behaviors can be measured either by recording the number of "ticks" heard during a given period of activity or by recording the time (from the clock) at the beginning and at the end of a particular behavioral act.

If the exercise is so designed that the entire class is using the same 10-minute video-tape, you may wish to compare your results with those of your classmates. Review the taped sequence to verify individual interpretations of an action. Using a prepared data sheet which lists a few selected behaviors already identified, replay a selected portion of the tape and tally the frequency of their occurrence; replay it again and record the duration of one or two behavioral acts. Compare your data sheets with those of other students to get an estimation of your reliability

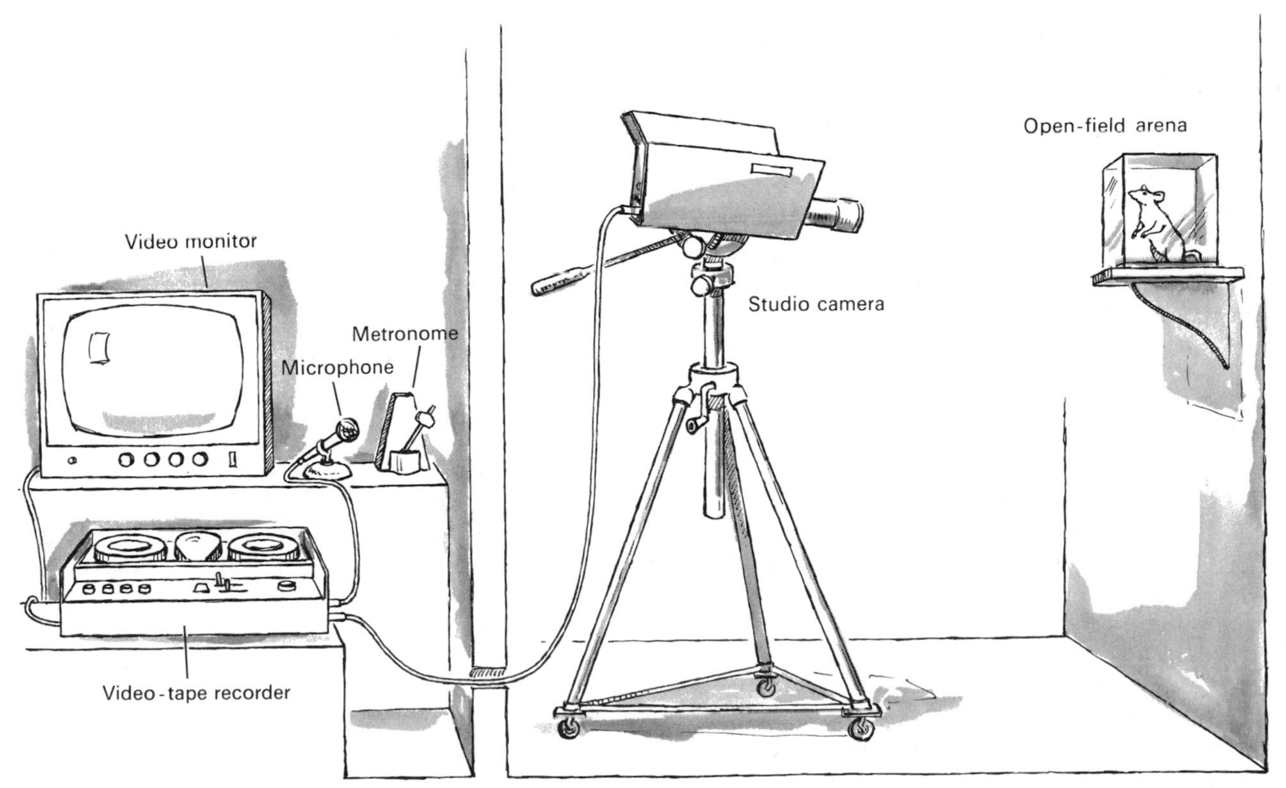

Figure 5-1. Diagrammatic set-up for using video-tape.

as an observer. You may also wish to subject your samples to an appropriate statistical analysis (e.g., analysis of variance).

QUESTIONS

1. How accurate are your identifications of the behavioral patterns observed? Are yours consistent with those of your classmates? How did your last observation compare with your first?
2. How discrete are these behavioral acts? When does one particular behavior end and another begin? Did any behaviors overlap or occur simultaneously? How might you account for this?
3. Did the instant replay of the original behavioral sequence enable you to refine your observations? In what ways?
4. Does the most frequently occurring behavior also have the longest average duration? What do your findings imply?
5. How would you construct a time-budget of the animal's activities? How could you use these data to study frequency of occurrence for these same behavioral acts?
6. How would you utilize this video-tape technique to study such complex behaviors as intention movements, displacement activity, locomotion, social interactions?

ADDITIONAL STUDIES

You have identified behavior and quantified it into temporal patterns. By drawing a grid on the floor of the open field (see Exercise 2) and by viewing the animals from above, you can quantify their behavior spatially. Social behavior can also be studied in more detail since the video-tape can be rerun as often as needed to break down the complex social interactions into quantifiable segments. You can search for consistent behavioral postures and patterns in studies dealing with the development of behavior. By recording vocalizations of the animal while taping, you may be able to correlate specific sounds with particular behavioral acts. A critical analysis of energy budgets of animals can be performed, even on a 24-hour basis, by setting up a random sampling schedule via an automatic timing device. This can also apply to such laboratory studies as orientation, homing, activity patterns, and so on. Shy or nocturnal animals can also be studied more easily with a video-tape system.

REFERENCES

Banks, E. M., 1962. A time and motion study of prefighting behavior in mice. *J. Genet. Psychol.* 101:165–183.

Barnett, S. A., 1963. *A study in behaviour: principles of ethology and behavioural physiology, displayed mainly in the rat.* London: Methuen.

Carpenter, C. C., and G. Grubitz, III, 1961. Time-motion study of a lizard. *Ecology* 42:199–200.

Clark, L. H., and W. W. Schein, 1966. Activities associated with conflict behavior in mice. *Anim. Behav.* 14:44–99.

Grant, E. C., and J. H. Mackintosh, 1963. A comparison of the social postures of some common laboratory rodents. *Behaviour* 21:246–259.

Levy, M. H., and I. Weissman, 1971. The VTR set-up at Blythedale. *Educ. Television* 3(6):10–11.

Schneirla, T. C., 1950. The relationship between observation and experimentation in the field study of behavior. *Ann. N.Y. Acad. Sci.* 51:1022–1044.

LABORATORY STUDIES

PETER MARLER
The Rockefeller University

Behavior of Day-Old Domestic Chicks and Ducklings

This exercise is designed to illustrate a number of basic principles of behavior with the relatively simple repertoire of actions of young chicks. You will begin with a study of their vocalizations in which you will try to determine what external stimuli are responsible for producing some of the different calls. Then, as a further study of the stimulus control of behavior, you will perform an exercise comparing the pecking color preferences in chicks and ducklings.

METHODS

1. External Stimuli Affecting Calls of the Chick

Subjects and Materials

The chicks should be as young as possible, preferably hatched in an incubator in the laboratory. Each chick should be marked for individual identification (e.g., colored leg band). Do not attempt to sex the birds at this age.

Basic materials needed for each investigator are two small aquaria with lids, a stopwatch, some chick food, a frosted-glass plate and a glass rod for making sounds, and a pair of opaque glass jars large enough to hold a chick.

Procedure

A. Preliminary Observations Try to identify the different calls that the chicks utter, varying the situations in which the birds are placed. Attempt to distinguish two major classes, so-called "pleasure" calls and "distress" calls. It is desirable for each student or pair of students to work in some isolation to avoid mutual disturbance. The plan is to record the frequency of the two types of calls given under the various conditions listed below. Adopt a standard recording period such as one minute. Use a stop watch. Hand-operated click counters may help, especially with rapid sequences of pleasure calls. Record the data in tabular form and summarize before going on to the next part of the experiment.

B. Experimental Conditions Observe each bird for at least ten minutes under each of the following conditions (when two birds are together, record the activity of just one individual):

a. Two chicks together in the same aquarium.
b. Two chicks visible to each other but in some degree of acoustical isolation. Place the birds in separate aquaria side by side with a heavy lid on top. Make a hole in the lid over the bird you are studying and listen through it.
c. Food. Test individual chicks with and without food on the floor of the aquarium. You may wish to deprive the birds of food for a short period, say 15 or 30 minutes, before conducting this experiment.
d. Exposure to sounds from a group of chicks that cannot be seen. For this purpose cover the walls of an aquarium with an opaque material and place a chick inside it. Then place the aquarium near a group of chicks.

e. Exposure to sounds of different types. While the chick is still in the aquarium with opaque walls compare the effect of tapping a glass rod on a frosted glass plate in the rhythm of a clucking hen, about 4 per second, with the sound made by rubbing the tip of the rod from one end to the other of the plate, on the frosted side. The scraping sounds should be spaced out, 4–5 seconds apart, somewhat in the rhythm of the call of a cockerel at the approach of a flying predator.

f. Warmth and cold. Compare the behavior of a chick that is placed in an opaque jar that has been chilled in the refrigerator with its behavior when placed in a jar that has been heated in hot water.

Having completed this series, select the two factors with the most marked effects and expose a chick to both at the same time. Do the effects summate arithmetically or not?

2. Pecking Behavior of Chicks and Ducklings

What kinds of stimuli evoke pecking? We know that chicks and ducklings are predisposed to peck more at some colors than at others, and that there are specific differences in these color preferences.

METHODS

Subjects and Materials

For this exercise you will need both chicks and ducklings. Prepare beforehand a collection of plaster, plastic, or wooden beads or balls, fastened in pairs on the end of tongue depressors (Fig. 6.1). They can be colored with watercolor paints and then coated with transparent cement to protect them. Use four or five bright colors.

Procedure

The method is to present a pair of colored objects to a chick to see which it pecks first.

Figure 6-1. Apparatus for studying color preferences of young birds: color beads or balls on ends of a tongue depressor.

Perform this experiment with a large group of chicks, selecting the individuals which are predisposed to peck at the moment. Prepare the color combinations to permit tests of all possible pairs of the colors selected; you should have two sets of each combination, with the positions of the colors different on each. It may be useful to put the same pair of colors on each end of a tongue depressor with the positions reversed. The class can adopt some general program for administering these stimuli, perhaps by randomizing the order of stimulus presentation. A total of twenty choices should suffice to define the preference for each color pair. Sum your results and then combine those of the whole class.

Repeat the experiment with a group of ducklings.

The drawback of a simple experiment of this type is that it fails to account for effects of variations in hue, saturation, and so on, as different variables, which you would want to control in a more sophisticated experiment. However this method suffices to establish a set of preferences in chicks, which can then be compared with the preferences of ducklings. Speculate on the possible ecological significance of the differences in color preferences.

As a further exercise, measure the time a bird takes to decide which color to peck, and see whether this relates to the preference rating of the two colors. The difficulty lies in deciding when the stimulus patterns are first fixated. Certain individuals may seem to be making random choices. Does their pecking latency differ from that of birds which are comparing before choosing?

REFERENCES

Bermant, G., 1963. Intensity and role of distress calling in chicks as a function of social contact. *Anim. Behav.* 11:514–517.

Collias, N. E., 1952. The development of social behavior in birds. *Auk* 69:127–159.

Collias, N. E., and M. Joos, 1953. The spectrographic analysis of sound signals of the domestic fowl. *Behaviour* 5:175–188.

Hess, E. H., 1956. Natural preferences of chicks and ducklings for objects of different colors. *Psychol. Rep.* 2:477–483.

Kaufman, I. C., and R. A. Hinde, 1961. Factors influencing distress calling in chicks, with special reference to temperature changes and social isolation. *Anim. Behav.* 9:197–204.

PETER MARLER
The Rockefeller University

Effects of Male Hormone on the Behavior of Chicks

The effects of a change in physiological state on behavior can be demonstrated by treating domestic chicks with male sex hormone. The objective of this exercise is to study the changes in behavior of the chick that occur with age under normal development and the changes that occur in chicks that have been treated with testosterone propionate.

METHODS

Subjects and Materials

Chicks used should be as young as possible (see Exercise 6). Mark each bird for individual identification.

Male sex hormone is best administered as crystalline testosterone propionate, available from drug suppliers as a 75 mg cylindrical pellet. It can be cut with a razor blade into approximately 10 mg portions.

Procedure

A. First Week Place two chicks together in a cage with food, water, and a substrate such as sand, gravel, or shavings. Observe the social and maintenance behaviors exhibited. Describe and catalogue these patterns. You will use these notes during the next observation period to compare the rate of behavioral development of these chicks relative to those injected with hormone.

At the end of this first period, implant several of the birds with a pellet of testosterone propionate. Prepare pellets containing 10 mg of testosterone. Make a slit in the skin on the back of the neck (Fig. 7.1). Lift the skin with forceps. Insert the pellet about half a centimeter away from the incision toward the back of the head: if it is left too close to the incision it may be cast off with scar tissue. A drop of collodion (available from drug companies) suffices to seal the incision afterwards. An equal number of control birds should be subjected to the same procedure, but without a pellet being inserted. Compare the experimental and control birds one week later.

B. Second Week Place a hormone-treated and a control chick together in a cage. Birds should be individually marked for rapid identification. Compare the following traits:

1. Postural differences. Use the correct terms for designating feather tracts and parts of the body. Do not hesitate to use sketches and drawings.
2. Facial appearance (particularly around the eyes).
3. Manner of walking.
4. Voice. The crowing patterns of the implanted chicks may be recorded and subjected to sound spectrographic analysis.
5. Form and frequency of other motor patterns, such as wing flapping, stretching, yawning, and so on.

Make a list of the newly developed behavior patterns in the hormone-treated bird, and compare them with those of the control, taking account of the changes in the behavior of controls that have occurred in the past week. Are any of the pre-existing behavior patterns of the treated chick exaggerated, or have any been lost? Make a quantitative comparison of the frequency of some of the actions observed.

If time permits, conduct other social encounters between hormone-implanted birds and controls, in various combinations (different numbers of chicks), and look for any differences in their behavior.

Count the frequency of the various types of posturing, pecking, and so on. Try to distinguish pecking actions that are suggestive of fighting from those that are clearly nonaggressive.

QUESTIONS

1. What general conclusions can you draw about the effects of testosterone on chick behavior?
2. Is it important to know the sex of your chicks? (Note that day-old chicks supplied by dealers are very often all males.)

ADDITIONAL STUDIES

If time permits at the end of the second period, autopsy some experimental and control chicks and dissect out the testes. Try to explain any variation in testis size that you can detect.

An additional study requiring more time would be to compare the effect of male hormone on male chicks with its effect on female chicks. In addition, hormone could be administered to chicks of the same sex at different developmental ages to assess the effect of maturational stage on the response to exogenous testosterone treatment.

REFERENCE

Marler, P., M. Kreith, and E. Willis, 1960. An analysis of testosterone-induced crowing in young domestic cockerels. *Anim. Behav.* 10:48–54.

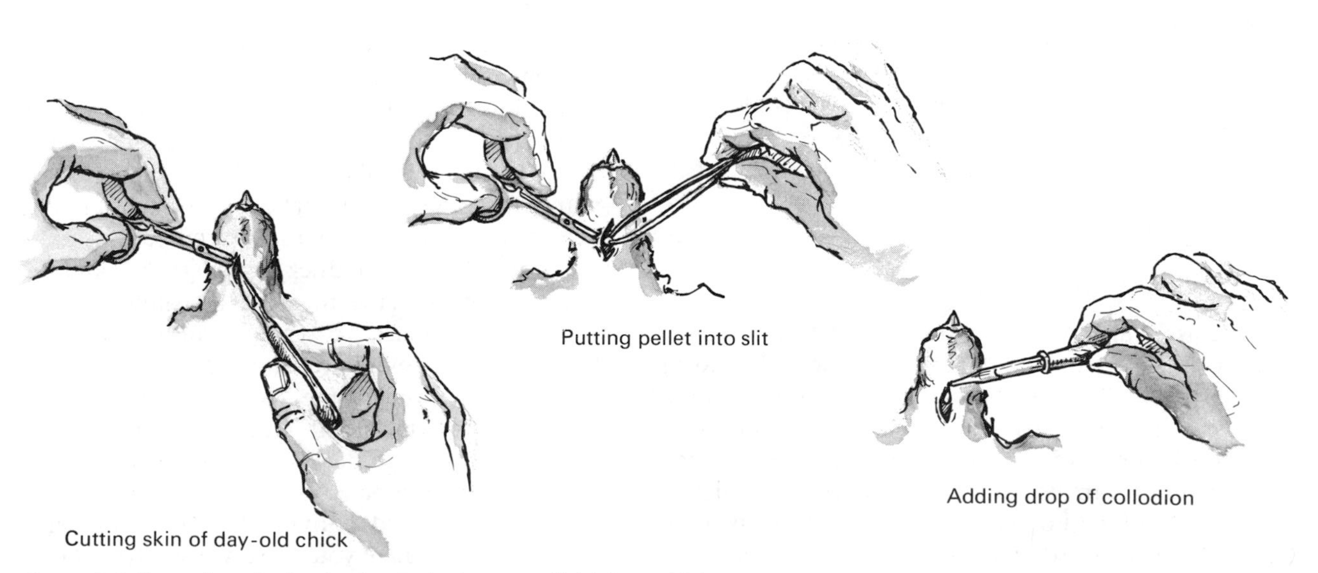

Figure 7-1. Procedure for implanting testosterone pellet into a chick.

PETER H. KLOPFER
Duke University

Imprinting

Newly hatched precocial birds (e.g., ducks, geese, chickens, turkeys) will generally follow a wide range of objects made to move before them. As a consequence of following some object during a period within 48 hours after hatching, the young bird's following response will subsequently persist for many days or weeks. However, if following is not elicited during this early period, it is believed that it generally cannot be elicited at a later age. Your purpose in this exercise is to determine the degree of specificity in the preference for the model originally followed. ("Imprinting" is the word used to refer to the persistence of a specific preference as a consequence of an exposure during a specific developmental period, the "critical" period.)

METHODS

Subjects and Materials

The class will have available at least twenty young birds (ducks, geese, or some other "heavy" kind of domestic fowl). These newly hatched birds should be kept in a brooder with separate compartments (approximately 15 × 10 × 10 cm) for each bird, so that they can be maintained in both visual and tactile isolation. For compartments, use painted or opaque plastic freezer boxes, or construct them out of cardboard and masking tape. Two kinds of training and testing apparatus are depicted in Figures 8.1 and 8.2 (the same apparatus is used for both training and testing).

The birds should be 16 ± 8 hours old (post-hatch) for maximum responsiveness. If you use developmental or incubation age, appropriate intervals would be 27 days and 12 hours to 28 days and 12 hours for ducklings; 21 days to 22 days for chicks.

Procedure

A. Training Divide your individually isolated birds into two populations and expose those of one population, one at a time, to a model (a life-size decoy, a cardboard figure, or something similar) suspended from a pulley-line which is made to move (manually, or with a motor) either around a circular arena or along an alley at a speed comparable to that of a normal chick or duckling (Figs. 8.1 and 8.2). If the model is made to move intermittently (e.g., 10-second movement, 5-second pause) and emit a short repetitive sound (you may use a human voice saying "*come*, come, come, come" or the clucks of a mother hen), which will require installation of a miniature tape recorder and a loudspeaker in the model, the proportion of "followers" should increase.

Expose each bird individually to the model for 20–30 minutes. Make your observations from behind a curtain or screen. Record the number of seconds

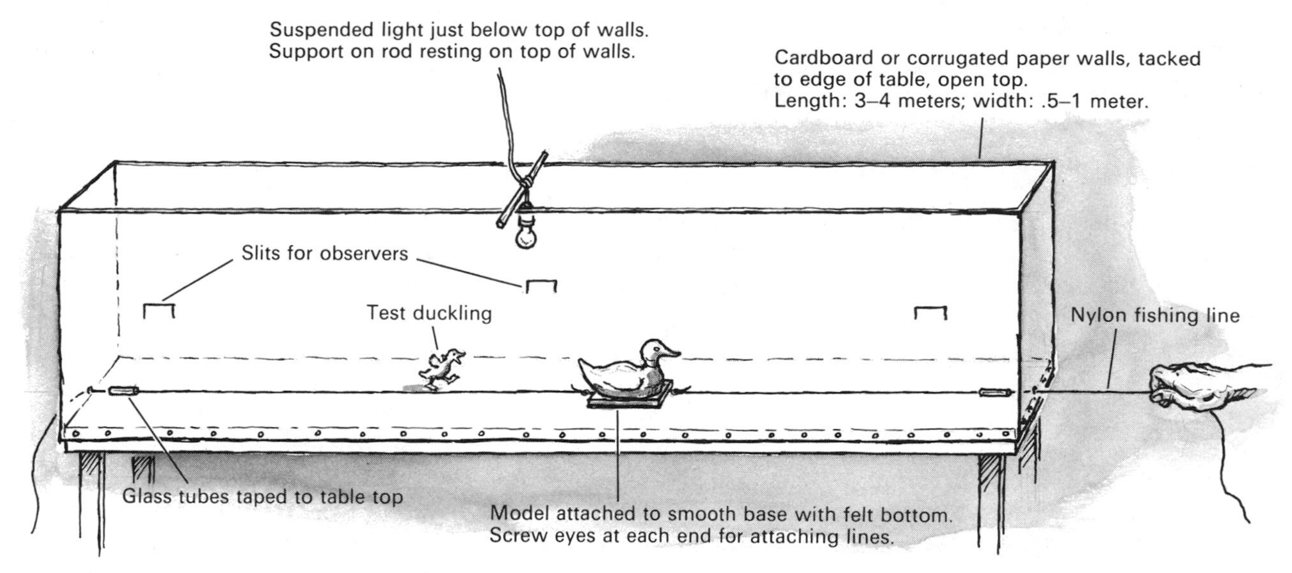

Figure 8-1. Alley-type imprinting apparatus.

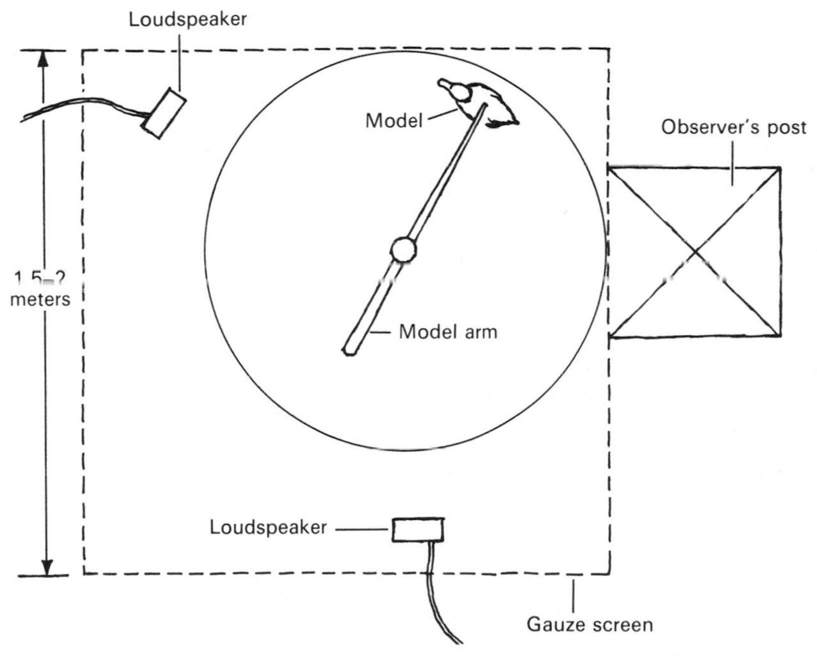

Figure 8-2. Circular-type imprinting apparatus. The model is rotated by a simple electric motor appropriately geared.

the bird spends following the model within a standard distance (e.g., 20 cm) from it.

Each bird from the second population should then be placed in the testing area for an identical period, but no model should be visible to the birds, and no sounds directed at them.

B. Testing After the training has been conducted, return the birds to their compartments and then retest both populations 24–26* hours later, at an age when they would ordinarily no longer be prone to follow *de novo*, i.e., after the peak of the critical period for the elicitation of following has passed. The "followers" will of course still respond, having now been "imprinted." But this time allow them to choose between the original model and some other

*Longer intervals are possible, but then the birds must be given light, food, and water.

simultaneously presented model which differs in one or more features (shape, color, size, speed of movement). In the circular apparatus (Fig. 8.2) the models can be hung from opposite arms of the T; in the rectangular apparatus they can be presented side by side.

Compare the responses of the birds from the two populations. Also compare the responses of those in the trained population that did not follow the model during training with those that did.

QUESTIONS

1. Is following during training a prerequisite to the maintenance of the following response?
2. Is exposure to a model during the training period necessary to assure following at a later age?
3. Are all models equally effective?
4. Is the original "imprinted" model always preferred?
5. What do your answers to questions 3 and 4 imply about the nature of imprinting?

REFERENCES

Gottlieb, G., 1963. Following-response in ducklings: age and sensory stimulation. *Science* 140:399–400.

Hess, E. H., 1958. "Imprinting" in animals. *Sci. Amer.* 198(3):81-90. (Offprint 416.)

Klopfer, P. H., 1965. Imprinting: a reassessment. *Science* 147:302–303.

Klopfer, P. H., 1973. *Behavioral aspects of ecology,* 2d ed. Englewood Cliffs, N.J.: Prentice-Hall.

Klopfer, P. H., 1973. Imprinting: monocular and binocular cues in object discrimination. *J. Comp. Physiol. Psychol.* 84:482–487.

Sluckin, W., 1964. *Imprinting and early learning,* Chicago: Aldine.

JOHN A. ROSS
St. Lawrence University

Feeding Behavior in the Blowfly

To survive in a given niche a species must evolve physiological and behavioral mechanisms that enable it to exploit appropriate energy sources efficiently. The feeding behavior of most complex organisms (e.g., mammals) is relatively flexible and subject to modification through experience. Many animals with less complex nervous systems (e.g., insects) have evolved more rigid or stereotyped responses to potential sources of energy. In this manner survival of the species is maximized.

Being an insect, the blowfly (*Sarcophaga*) has a rather simple nervous system and shows relatively stereotyped behavioral patterns. The niche of the blowfly is such that there exists a close correlation between the stimulative and nutritional properties of available food items. As a blowfly alights on a food source, several types of receptors on its feet are stimulated. If the activity in the "sweetness receptor," for example, is greater than the activity in the "salt-acid receptor," the blowfly will extend its proboscis to the substrate. If the balance of activity favors the latter receptor, it will withdraw. Stimulation of the oral papillae on the proboscis elicits ingestion until it is inhibited by activation of the stretch receptors in the foregut and the body wall. Thus, the fly survives.

The purpose of this study is to determine differences in the feeding behavior (frequency and duration of proboscis extension) of the adult blowfly when exposed to sugar solutions of different concentrations and chemical compositions.

METHODS

Subjects and Materials

Blowflies may be obtained from biological supply houses either as larvae or pupae (either stage is acceptable). Store the flies in an aquarium, insect cage, or a similar screen-covered box. While the pupae or larvae are developing it is essential to cover them with a damp cloth or damp vermiculite and place them in a warm area (27°C). Adults will develop in one or two weeks depending on the original stage. Feeding is unnecessary during all stages of development and would seriously interfere with this study. Approximately a hundred adult flies will suffice.

Any of several sugars, such as glucose, fructose, lactose, sucrose, or mannose can be used (available from chemical supply houses). The concentration of the chosen sugars should vary from zero to 5 percent. Each laboratory group will also need a stopwatch, a microscope slide, and an open-bottomed wire-screen cage approximately the same size as the slide and 2.5 cm high.

Procedure

Spread several drops of one of the sugar solutions on one end of the microscope slide, trap a fly under the wire-screen cage and lower it over the slide. Using a stopwatch, record the time (latency) between stepping into the solution and proboscis extension into the solution. Then, until the fly leaves the sugar solution, record (1) the frequency of proboscis extension, (2) the duration of each extension, and (3) the total time that the blowfly is in contact with the solution.

At least five different concentrations of two sugars should be tested and a minimum of five flies should be used for each solution. The microscope slide must be washed thoroughly with soap and water, or replaced, after each test.

The pooled results of this study should be plotted in the form of several bar graphs: one each for mean scores for latency, number of proboscis extensions, duration of extensions, and total time in contact with the solution. Inspection of these graphs will indicate differences in the response of the flies to the various sugars and concentrations used. The data may be statistically analyzed by analysis-of-variance (anova) procedures to determine whether there is a differential response to increasing concentrations. (See Edwards, 1958.)

DISCUSSION

An understanding of the results presupposes an understanding of the general behavior of the blowfly and the anatomical control of this behavior. This insect has a simple nervous system and stepping into a sweet substrate serves as a stimulus for proboscis extension. The latency of extension provides a means of measuring the relative stimulative value of the solution being tested and constitutes the resultant effect of tendencies for extension and withdrawal from the solution.

Consult Dethier and Stellar (1970) and Gelperin (1966) for more information on blowfly feeding behavior.

QUESTIONS

1. How could a system evolve that cannot distinguish between nutritive and nonnutritive substances?
2. Assuming that the stretch receptors terminate ingestion, how can the fly in its natural environment regulate its caloric intake?
3. Did you find the blowfly responding in an identical manner each time the stimulus was presented, or was there some variability in behavior? Do you have an explanation?

ADDITIONAL STUDIES

An alternative procedure would be to give each fly several sugars or several concentrations on the same slide at the same time to determine preferences. Do the flies ingest more of one solution than the other? Use the same dependent measures as before.

A second alternative would be to have the fly step in one solution while its proboscis extends into another. This can be accomplished by gluing a toothpick or small stick on the fly's back, using a rapid drying cement or nail polish, then lowering the fly's feet into one solution, making sure that the proboscis extends into a nearby drop of another solution. Again, record the same data. Do these results agree with previous ones? (Note: The gluing procedure can be facilitated by rendering the fly inactive by placing it in a refrigerator for five minutes.)

A third alternative is to measure the volume of solution the fly ingests. In this case a 16 mm (inside diameter) glass tube should be bent into a J shape at its tip. The tip must end in a paraffin dish from which the fly drinks with the tube acting as a reservoir to keep the dish filled (Fig. 9.1). The volume of the solution ingested can be determined by marking the tube at the beginning of the test, refilling it at the end of the laboratory period (using a 1 ml or smaller syringe), and recording the amount of solution used.

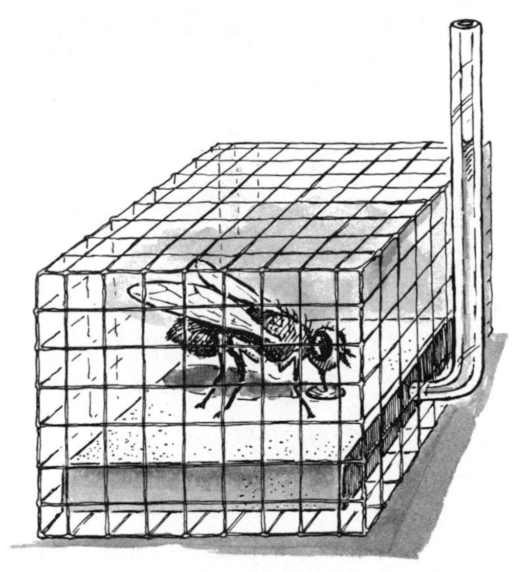

Figure 9-1. Blowfly ingesting from paraffin dish.

REFERENCES

Dethier, V. G., 1962. *To know a fly.* San Francisco: Holden-Day.

Dethier, V. G., 1966. Insects and the concept of motivation. *In* D. Lewine, Ed., *Nebraska symposium on motivation,* pp. 105–136. Lincoln: University of Nebraska Press.

Dethier, V. G., and E. Stellar, 1970. *Animal behavior.* Englewood Cliffs, N.J.: Prentice-Hall.

Edwards, A. L., 1958. *Experimental design in psychological research.* New York: Rinehart.

Gelperin, A., 1966. Investigations of a foregut receptor essential to taste threshold regulation in the blowfly. *J. Insect Physiol.* 12:829–841.

Gelperin, A., and V. G. Dethier, 1967. Long-term regulation of sugar intake by the blowfly. *Physiol. Zool.* 40:218–228.

GORDON M. BURGHARDT
University of Tennessee

Sensory Cues and Feeding Behavior of Snakes

Feeding is one of the most universal and frequent behaviors engaged in by all species of animals. Ethologists have discovered that different species not only differ in their sensitivity to different types of sensory information (e.g., colors, odors, sounds) associated with potential food items but that many animals are selective to particular relevant cues. Stimuli that elicit specific behavior patterns are termed sign or key stimuli. Sign stimuli emanating from conspecifics and used in social interactions are called releasers. The responses to sign stimuli are influenced by the immediate environment of the animal as well as its motivational state. Feeding behavior readily demonstrates the operation of sensory cues in behavior since it is frequently exhibited in captivity by most species and motivation can be altered by food deprivation or satiation.

Snakes possess many unique adaptations for feeding, some of which will become very evident to you. Besides vision and olfaction, the forked tongue of snakes is associated with a chemosensory organ called the Jacobson's or vomeronasal organ. Gustation has been little studied but seems of little importance until the snake's prey is in its mouth since no taste buds have been found on the tongue. Snakes hear airborne sounds poorly at best; most workers claim that snakes are deaf. Snakes are very sensitive to vibratory and tactile cues, however.

Snakes are one of the few groups of animals in which every known member is an exclusive carnivore. The large meals that snakes consume and their ability to fast for long periods have inhibited experimental work. The present exercise focuses on feeding behavior in snakes. It emphasizes observing how snakes feed, and how the sensory control of their feeding relates to the ethological concept of sign stimuli, most studies of which have utilized visual cues.

METHODS

Subjects and Materials

Garter snakes of the genus *Thamnophis*, the most suitable snakes for this experiment, are readily and inexpensively available from reptile dealers and biological supply houses or can be captured locally. Perhaps the best species to use is the widespread common or Eastern garter snake, *Thamnophis sirtalis,* which lives in a variety of habitats including fields and open woodland, often near water. This species eats earthworms, fish, frogs, tadpoles, salamanders, and leeches. Obtain some of these prey species, along with a few small lab mice and such insects as grasshoppers, crickets, or mealworms, all of which are rarely or never eaten. Keep the prey in a different room from the snakes. If this is not practical, keep them on opposite sides of a large room.

House the snakes individually in wooden cages with one glass side, or in aquaria with removable screen escapeproof lids and having 1000–3000 cm^2

floor space. Visually isolate the snakes from one another, using partitions if necessary. If the exercise is conducted in summer or early fall, some of your snakes may be pregnant and give birth to litters of 15 to 75 young. House the young separately or in very small groups. Plastic mouse cages with filter tops are ideal. If possible use the young in the feeding studies described below while they are still naive. They should be fed small pieces of earthworm or guppy-sized fish when they are 5–15 days old.

Line each cage with semiabsorbent paper (newspaper will do) and provide a water dish heavy enough not to be easily overturned and large enough for the animal to coil up in. Also provide a rough stone or brick to aid shedding and a piece of bark or folded section of cardboard for the animal to hide under; these may be removed during testing. Temperature should be maintained at about 25°C although cooler temperatures at night are permissible. If a snake's eyes become cloudy, it is a sign that it will shed in a few days. However, the eyes may clear up for a day or two immediately before shedding. Although feeding behavior may be inhibited at this time, carry through the testing and compare your results with those from snakes that are not shedding.

Useful items for this exercise are short and long forceps, glass vials, a hot plate, beakers, distilled water, a pan balance, cotton swabs at least 15 cm long or home-made equivalents, a stopwatch, and a hand counter.

When working with snakes it is best to test them in their home environment with a minimum of disturbance. Maintaining the snakes in a small room where they are not disturbed for at least an hour before testing or observing is ideal. Figure 10.1 illustrates a possible laboratory set-up for testing snakes. Vibrations can disturb snakes, but soft voices apparently do not. Handling the snakes, particularly juveniles, is generally not advisable.

Figure 10-1. Laboratory set-up for testing snakes. Since the light under the table is the only light in the room, the experimenter is in the shadow. (Based on a figure by Doris Grove.)

Procedure

Ideally, the snakes will have been maintained in their quarters for several weeks before observation and have eaten regularly every 3–5 days: earthworms and minnows are a good convenient diet. They should not be fed for 4–7 days before the exercise begins. If enough snakes are available, divide them into two groups, one group for general observations and one for a more in-depth investigation of sensory cues involved in feeding.

A. General Observations First, observe the snakes when they are fed various foods. With forceps introduce a fish or earthworm in front of the animal's snout. Note and describe such behaviors as tongue flicking, orienting, striking, grasping, manipulating or moving around the prey, and ingesting. Try different sizes of prey including some with a diameter larger than the snake's head. Similarly present live prey such as insects or small mice, which these snakes do not normally eat. Count and record the number of tongue flicks and note the latencies of the various behaviors. Is the prey swallowed headfirst or tailfirst? Does the tongue seem to contact the prey before a strike?

Second, place both live and dead (killed by immersion in hot water) prey (e.g., minnows, earthworms) in various locations in the cage and record the various behaviors exhibited in locating and ingesting the prey. Placing fish in the water dish, for example, usually yields good results. Draw some tentative conclusions about the role of various cues emanating from the prey.

If the snakes are near satiation, and if unfed snakes are not available, cease testing and wait 4–6 days before beginning the second part of the exercise. Snakes, unlike most animals, do not normally eat every day, a characteristic you should appreciate from observing the size and amount of prey they will ingest at one feeding. If enough snakes are available try feeding the animals you have been testing to satiation each day. Offer minnows or worms at five-minute intervals repeatedly until food is rejected twice in a row (use 60-second trials). Determine the cyclical nature of snake feeding in this manner. (See also Myer and Kowell, 1971.)

B. Sensory Cues After gaining some acquaintance with the feeding behavior of snakes, you are now ready to determine the role of visual and chemical stimuli in predatory behavior. You will concentrate upon the stimuli that release the initial orientation and attack on prey. Be alert to the occurrence of agonistic attacks. They can be differentiated from predatory attacks.

a. Place a live worm in a glass beaker on an elevated platform outside the glass side of the snake's cage. Note the reaction and the frequency and latency of any behaviors such as approach and tongue flicking. Remove after 3 minutes. Repeat with a nonmoving dead worm. Do the same with live and dead fish in water and a small mouse or insect. To control for odors, cover the prey containers. Then place each prey in a glass vial, seal the top, and place it in the snake's cage. What can you conclude about the role of visual cues such as movement, pattern, form, and size? Which seems most important? Why?

b. With forceps offer the snake a dead worm or fish from outside the glass side, allowing any odors from the prey to enter the cage. Move it back and forth. Note any orientation, tongue flicking, prey following, or attacking. Then, offer the snake the prey inside the cage and observe similarities and differences in response. Did the snake attempt to attack the prey through the glass? If it attacked the prey in the cage but made no attempt to do so through the glass what was the determining factor (odors and visual cues were present in both cases)?

c. Take a cotton swab, dip it into water, shake off the excess, and slowly bring it within about 1–2 cm of the snake's snout. Hold it there for 60 seconds and note the snake's behavior, especially orientation and tongue-flick frequency. Repeat this procedure periodically throughout the next step.

d. Take another swab, dip it into the water, rub it thoroughly on the side of a moist earthworm. Then present this swab in the same manner as in step *c.* Observe the reactions. If the swab leads to an open-mouth attack, remove the swab before the snake's teeth get entangled in it. It may take a few tests before you become adept at this. Repeat the tests using other kinds of prey the snake will eat (fish, frogs) and at least one it will not attack (mouse or insect). If several snakes are available balance the order of testing between individuals.

For more systematic tests, prepare chemical extracts by placing a clean live prey animal in 60°C distilled water for two minutes in the ratio of 3g prey/10 ml H_2O. Centrifuge or filter the liquid and place in sealed glass vials. It may be refrigerated for up to three days or frozen indefinitely.

Compare the behavior of individual snakes and their responses to different prey stimuli in steps *a* to *d.* Subject your data to statistical analysis, if feasible.

QUESTIONS

1. What conclusions can you draw about feeding behavior and its sensory control in garter snakes?
2. What seem to be the respective roles of the tongue–Jacobson's organ system (gustation can be largely ignored for attack behavior)–olfaction, and visual cues in the attack on prey? Can these sensory modalities be linked in the same way to the arousal of interest in, and orientation to, prey stimuli?
3. What further experimental controls and tests would be useful in confirming or extending your conclusions?
4. How might your findings relate to the ecology and evolution of the species as well as to the individual life history of the subjects you tested?
5. Are some stimuli and foods recorded as being eaten by this species more effective than others?
6. Does the effectiveness of chemical cues justify considering them as sign stimuli for attack behavior?

ADDITIONAL STUDIES

If animals, equipment, and time are available, the following problems may be also investigated.

1. Test other species of snakes with different feeding habits and develop a comparative study to gain information on species differences. The extract tests might not be effective with species from some genera of snakes. Other than the possibility that chemical cues may be less important in these species, a snake's temperament and the testing method used may affect your results. Although this is true for many of the rodent-eating snakes (e.g., *Boa, Elaphe*)–which, unlike garter snakes, are constrictors–observing the capture and ingestion of a mouse by one of these forms can be particularly instructive, since the prey are shaped more irregularly and can put up a stronger defense. The third major feeding method involves the injection of a deadly or paralytic poison as by vipers or cobras, species which are unlikely to be utilized in beginning laboratory studies. Prey size is important with these snakes (Loop and Bailey, 1972).
2. If newborn or newly hatched and previously unfed snakes are available, by all means isolate and test them in the manner above. If a fairly large litter is available you may want to systematize your extract preparation, testing, and recording procedures as suggested in Burghardt (1969) and other papers. Modify these methods to suit your purposes. Are the results congruent with the observations upon which the classical ethological concept of the Innate Releasing Mechanism (IRM) is based?
3. What might be the role of feeding experience on the prey preferences shown by snakes? If newborn snakes show the same preferences as adults does this rule out any modification of eating habits? How would you test for changes in prey preferences, using a species which eats two or more kinds of prey that are readily available for testing?

REFERENCES

Bellairs, A., 1970. *The life of reptiles.* New York: Universe Books. 2 vol.

Burghardt, G. M., 1966. Stimulus control of the prey attack response in naive garter snakes. *Psychonomic Sci. Sect. Anim. Physiol. Psychol.* 4:37–38.

Burghardt, G. M., 1969. Comparative prey-attack studies in newborn snakes of the genus *Thamnophis. Behaviour* 33:77–114.

Burghardt, G. M., 1970. Chemical perception in reptiles. *In* J. W. Johnston, Jr., D. G. Moulton, and A. Turk (Eds.), *Communication by chemical signals,* pp. 241–308. New York: Appleton-Century-Crofts.

Carpenter, C. C., 1952. Comparative ecology of the common garter snake (*Thamnophis s. sirtalis*), the ribbon snake (*Thamnophis s. sauritus*), and Butler's garter snake (*Thamnophis butleri*) in mixed populations. *Ecol. Monogr.* 22:235–258.

Conant, R., 1958. *A field guide to reptiles and amphibians of North America east of the 100th meridian.* Boston: Houghton-Mifflin.

Gans, C., 1961. The feeding mechanism of snakes and its possible evolution. *Amer. Zool.* 1:217–227.

Loop, M. S., and L. G. Bailey, 1972. The effect of relative prey size on the ingestion behavior of rodent-eating snakes. *Psychonomic Sci. Sect. Anim. Physiol. Psychol.* 28:167–169.

Myer, J. S., and A. P. Kowell, 1971. Eating patterns and body weight change of snakes when eating and when food deprived. *Physiol. Behav.* 6:71–74.

Stebbins, R. C., 1966. *A field guide to western reptiles and amphibians.* Boston: Houghton-Mifflin.

11

STANLEY C. RATNER
Michigan State University
LOUIS E. GARDNER
Creighton University

Habituation in Earthworms

The name "earthworm" refers to a number of different species of annelids, any one of which may be used in the following laboratory studies. The night crawler, *Lumbricus terrestris,* is a convenient species to use because individuals are large and their movements are conspicuous, and in addition is the species most often described in biology texts. However, you might find and use specimens of the garden worm (genus *Allolobophora*) or the red garden worm, also called the manure worm, (genus *Eisenia*). Worms of these groups are smaller than the night crawler, but the behaviors of all are very similar.

In this exercise you will study the responses of earthworms to the stimulus of vibration, and changes in their responses with repeated stimulation. It serves as a demonstration of habituation, a simple form of learning found in almost all animal species. Habituation in this context consists of learning not to respond to a certain stimulus as a result of repeated presentations of that stimulus, and hence must be distinguished from muscular fatigue and sensory adaptation. The process of habituation and its characteristics are discussed by Hinde (1970) and Denny and Ratner (1970).

METHODS

Subjects and Materials

Two worms are needed for each test series. One worm is tested; the other is observed as a control animal. Two recommended experimental apparatuses are shown in Figure 11.1. The preferred apparatus (Fig. 11.1,a) consists of a clear plastic tube (e.g., a piece of Tygon tube) 0.5 to 1.0 cm inside diameter with many small holes drilled or burned along its top. The worm is allowed to crawl into the tube and the tube is taped together to form a circular pathway around which the worm can crawl. The bell buzzer that is screwed to the base on which the tube is placed should be connected to batteries or any other 6-volt source. Activation of the bell buzzer, causing the tube and base to vibrate, serves as the stimulus for the worm's response.

An alternative apparatus (Fig. 11.1,b) consists of a plastic dish or circular glass such as a Petri dish with a top. The worm is placed inside and the dish is struck with any object that causes a brief vibration: the worm can move anywhere in the dish before, during, and after each presentation of the stimulus.

Procedure

Experimentally naive earthworms should be placed in the dimly illuminated tubes or dishes at least ten minutes before testing. Worms are considered acclimated to the apparatus when they become relatively inactive.

After the worm has become acclimated to the test apparatus apply stimulation (vibration). Vibration can be produced by briefly activating the buzzer on the platform on which the tube is placed or by striking the dish containing the worm (dropping a wooden stick or lightweight mallet several inches onto the center of the top of the dish). How does the worm respond? Repeat the stimulation after a period of 10–20 seconds and again note the worm's response.

Once you have become familiar with the worm's response to vibration, you are ready to study the habituation phenomenon. Successive stimulus presentations should be as constant as possible (i.e., same duration and intensity). The stimulus should be administered every 10 to 20 seconds until a total of fifty presentations are made or until the worm stops responding for at least five successive trials (periods of stimulation). For each trial, record whether or not the worm responded to stimulation and note any additional observations that may be important for the interpretation of the results (e.g., doors slamming when the stimuli are presented, unusual movements of the worm, or errors in procedure caused by the experimenter). Note also any changes in the intensity of the responses.

You should also record the responses of the control worm. This worm is observed on the same schedule as the worm that is tested, but this worm is not stimulated in any special way. The data from this worm serve as an estimate of the basic movement pattern of a worm in the apparatus.

Compare the number of responses for the tested worm to the number of responses for the control worm. Then, note any changes in the number of responses after differing amounts of testing by considering the fifty trials in groups of ten. The number of observable responses is counted for each group of ten trials and the percentage of observable responses per group of trials is calculated (e.g., if the tested worm made eight observable responses in a group of ten trials, the percentage is 80). Teams of experimenters can combine the results of their observations to produce average scores for both experimental and control subjects.

a

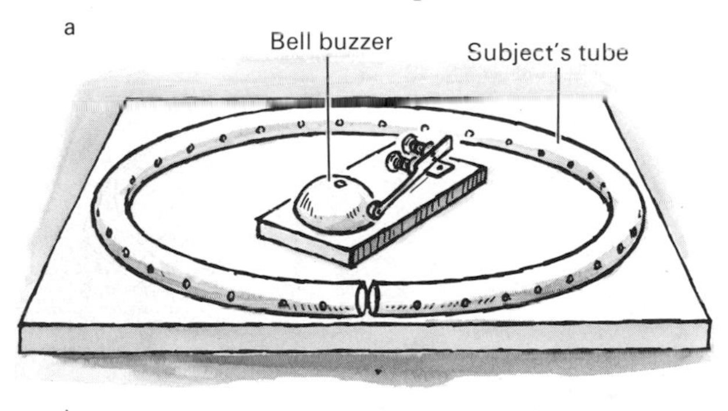

b

Figure 11-1. (*a*) Testing tube for studying habituation in worms. The clear plastic tube is vented with holes and mounted on a wooden base. Worm is placed inside the tube. (*b*) Alternative testing apparatus, a clear glass or plastic Petri dish with lid. Vibration is produced by tapping the lid.

QUESTIONS

1. Did the frequency of the response to vibration change with repeated stimulation?
2. How did the magnitude (intensity) of the response to vibration change with repeated stimulation? Do any such changes reflect a characteristic of habituation?
3. Is it likely that animals of other species show habituation to stimuli? Can you think of any examples of habituation in your own life or the lives of pets? Can taming an animal represent a case of habituation in which a complex stimulus or set of stimuli are employed?

ADDITIONAL STUDIES

Fatigue and adaptation. Habituation differs from muscular fatigue and sensory adaptation, although all of these processes can lead to weakening or reduction of responses to stimuli. Behavioral tests can be made after tests for habituation to determine if fatigue or adaptation are present. If some stimulus other than vibration can elicit the same response, then we assume it is not fatigued. Worms can be tested for fatigue by pinching them or blowing air on them. If some intensity or duration of vibration other than the one used to produce habituation can elicit the same response, then we assume the sensory system is not adapted.

Memory. How permanent is the habituation of the response to vibration? You can check your answer by testing the same worm several hours or one or two days after its first experience with another fifty trials. If the worm remembers, then a test using the same stimulus would show that only a few trials are necessary in order to habituate the response to vibration again. Denny and Ratner (1970) discuss evidence for the retention of the habituated response.

REFERENCES

Dales, R. P., 1963. *Annelids.* London: Hutchinson.

Denny, M. R., and S. C. Ratner, 1970. *Comparative psychology,* rev. ed., Homewood, Illinois: Dorsey Press. (See especially Chapter 5 on habituation.)

Edwards, C. A., and J. R. Lofty, 1973. *Biology of earthworms.* London: Chapman and Hall.

Hinde, R. A., 1970. *Animal behavior.* A synthesis of ethology and comparative psychology, 2d ed. New York: McGraw-Hill. (See especially Chapter 13 on habituation.)

Laverack, M. W., 1963. *The physiology of earthworms.* New York: Pergamon Press.

MARTIN W. SCHEIN
West Virginia University

12

Dust Bathing in Birds

This exercise utilizes one type of maintenance behavior, "dust bathing," to illustrate two major theoretical problems in animal behavior. The first concerns the role of experience in the expression of stereotyped behavior patterns. The second concerns changes in responsiveness under constant stimulus conditions, i.e., changes in the quantitative expression of the behavior over both short and long periods of time.

Recent studies suggest that dust bathing might be a component of thermoregulatory behavior: dust particles absorb excess oil on the feathers (particularly at the base) and therefore affect the insulating properties of the coat (Borchelt, et al, 1973; Healy and Thomas, 1973). Dust bathing is normally observed in birds on aggregate surfaces, such as loose soil, sawdust, and so on. Therefore it may be assumed that "particulateness" and "solidness" (as opposed to the "openness" of a wire-mesh surface) are important features of the stimuli eliciting the behavior. Other environmental considerations, such as temperature, humidity, and color of the particulate surface apparently have little effect on the quantity and quality of the behavior exhibited (Benson and Schein, 1965).

METHODS

Subjects and Materials

Domestic fowl, quail, and other gallinaceous birds commonly dust bathe and are suitable for study. (Japanese quail are especially suitable in view of their small size and ready adaptation to laboratory manipulations.) Half of the birds should be maintained for one or two weeks before the experiment in a cage or pen with any particulate animal bedding (such as sawdust, wood chips, ground corncobs, etc.) on the floor. These birds will have had an opportunity to dust bathe at will, and presumably will have had dust-bathing experience. The other half should be reared from hatching in cages with wire-mesh floors, so that they will have had no experience with solid floors.

Procedure

Experimental groups of 3–5 birds each should be subjected to one of six conditions, as follows:

Rearing Experience	*Test Condition*
Experience with bedding materials	Bedding floor
	Wire-mesh floor
	Solid (paper) floor
Wire reared; no experience with bedding materials	Bedding floor
	Wire-mesh floor
	Solid (paper) floor

The bedding-to-bedding and wire-to-wire groups serve as controls for each of the other test conditions within the experience category. The solid (paper) floor permits a distinction to be made between "particulateness" (bedding) and "nonparticulateness" (paper). It also permits a distinction between "solidness" (paper) and "openness" (wire-mesh). If possible, colors, temperatures of the floor surface, and so forth should be matched—or do these factors make any difference?

The test situation and presence of observers should not be permitted to become unduly alarming to the birds, since maintenance behaviors (grooming, preening, dusting) are easily masked by fear responses and escape attempts. At the very least, observers should move slowly and be generally unobtrusive; it may be necessary to construct simple blinds or to devise subterfuge systems for observation.

Each group of birds should be observed for at least 45 minutes with qualitative and quantitative aspects of dust bathing being noted. To gain as much information as possible, the class should be divided into observation teams of three persons each, and each team should assign responsibilities as follows.

Observer 1. Records the intensity of dust bathing, the amount of time spent dusting in comparison with time spent in other activities clearly unrelated to dusting.

Observer 2. Records the frequency of dust bathing, the number of birds participating in dusting activity at one-minute intervals.

Observer 3. Describes the pattern of dust bathing behavior. This observer identifies not only the components of dust bathing but also the sequence in which these components normally occur.

If time is available, a comparison with dust bathing in a second species should be made.

At the end of the observation period the data from all teams should be pooled so that statistical comparisons between treatment groups can be made.

QUESTIONS

1. Can you come to some agreement with the rest of the class on a definition of dust bathing? Where does it "fit" in the animal's repertoire of activities?
2. Can you agree on the components of dust bathing?
3. What stimuli seem to contribute to dust bathing?
4. To what extent does experience modify dust bathing motor patterns? Does it increase or decrease their intensity or duration?
5. In what ways do the intensity and pattern of dust bathing depend upon the test conditions rather than rearing experience?
6. To what extent is dust bathing contagious? Would this be adaptive?
7. What other functions might dust bathing serve besides regulation of body surface oil? Is there any social significance to the behavior?

ADDITIONAL STUDIES

Other variables that might be investigated are the effects of sex and age on dust-bathing behavior.

The effect of the availability of oil (from the oil gland at the base of the tail) on the quantitative and qualitative expression of dust bathing can be investigated by surgically removing the oil gland (the major source of feather oil). Oil is normally transferred from gland to feathers by preening.

The external stimuli releasing dusting behavior can be explored in several ways. You can use an overhead mirror to reflect different visual substrates onto a mirror floor. Alternatively, a sheet of clean glass over various substrates (wire, paper, sawdust, etc.) can be used to control tactile inputs while permitting modification of visual signals.

REFERENCES

Benson, B. N., and M. W. Schein, 1965. Factors influencing dust bathing in *Coturnix* quail. *Amer. Zool.* 5(2):196. (Abstract.)

Borchelt, P. L., J. Eyer, and D. S. McHenry, Jr., 1973. Dust bathing in bobwhite quail (*Colinus virginianus*) as a function of dust deprivation. *Behav. Biol.* 8(1):109–114.

Healy, W. M., and J. W. Thomas, 1973. Effects of dusting on plumage of Japanese quail. *Wilson Bull.* 85(4):442–448.

LOWELL L. BRUBAKER
Tennessee Wesleyan College

13

Wall-Seeking Behavior in Mice

Some animal species are generally found in open habitat; others occupy environments with more cover. Anyone wanting to trap mice should set his trap close to a wall since these animals tend to restrict their activity to sheltered areas. In fact, students of animal behavior have for a long time spoken of the wall-seeking or wall-hugging tendency of these rodents. This exercise will demonstrate some aspects of this behavior. It also shows that an ostensibly simple behavior raises many questions, and that these questions can be explored in a controlled laboratory environment.

The phrase "wall-seeking behavior" does more than describe the behavior; it interprets it. It says in effect that walls "attract" mice, that walls cause the observed nonrandom distribution of movement of mice placed in a novel environment. This implied hypothesis can be tested by varying the alleged cause of nonrandom movement. By comparing the behavior of mice subjected either to wall (W) or to no wall (NW) conditions you can determine whether walls are really necessary to elicit the nonrandom distribution of movement characterized as wall-seeking behavior.

METHODS

Subjects and Materials

A dimly lighted test arena will be needed for this study (Fig. 13.1). Although the size of the arena is not critical, it will be assumed in the discussion that the arena being used has a floor measuring 80 × 80 cm. The walls of the arena should be at least 30 cm high and must be removable to provide an unbounded field when required by the experiment. This feature also permits easy cleaning of the floor after each trial. Support the arena on an elevated surface, such as a table, that has smaller dimensions than the floor, so that the perimeter of the field provides a true edge and not just a "step down" to an even larger area. Mice will not jump off this surface when the walls are removed. A grid should be painted on the

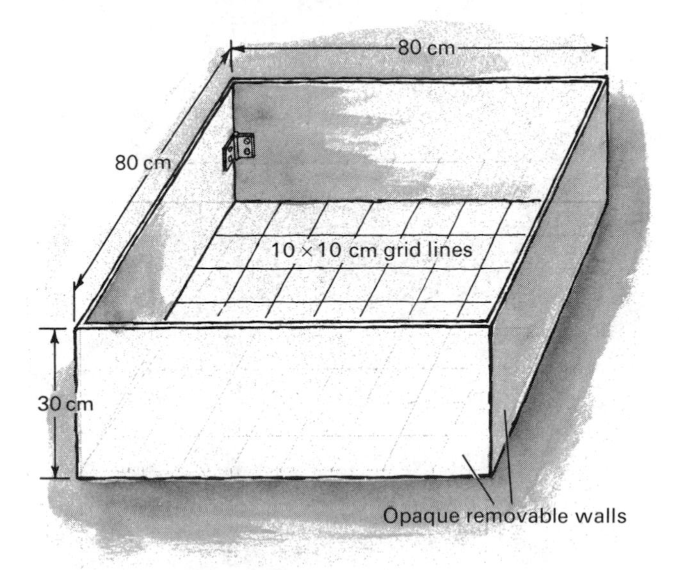

Figure 13-1. Wall condition (W).

floor to subdivide the area into sixty-four 10-cm squares.

Procedure

A. Response to Peripheral Walls The class should be divided into two groups. Both groups will follow the same procedure except that one will use the W (wall present) arena while the second will use the NW (no wall) arena (Fig. 13.2).

Each group of students should use at least ten mice. Test each animal individually in the arena for five minutes. Start half of the animals in the middle of the arena and half close to the wall (or edge). Every two seconds, as marked by the sound of a metronome beat or a tape-recorded beep, note on a data sheet representing the 64-square grid the position of the animal on the grid floor. Thus, during a five-minute test period, you will record 150 observations. At the end of this period remove the mouse, clean the floor, and introduce another test animal.

The total number of observations of the test animal in the 28 peripheral squares constitutes its W or NW score. Compute the means and standard errors for the animals in each of the two groups (Snedecor and Cochran, 1967, Chap. 2). How do the data of the two groups compare?

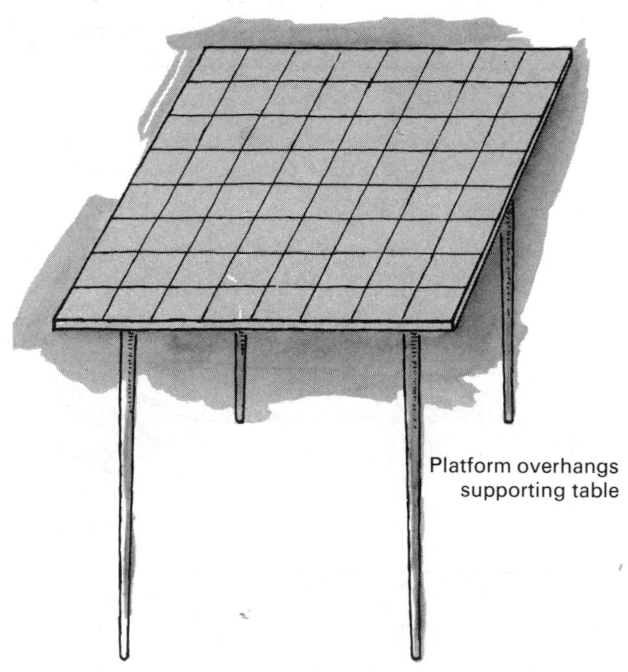

Figure 13-2. No wall condition (NW).

As a class, discuss the results of the experiments in terms of the differences and similarities of the stimulus conditions used. If the mice tested under both W and NW conditions spent more time at the periphery of the field than in the center, the wall-seeking hypothesis may not be an adequate explanation of the observed behavior. What alternative hypotheses can you think of?

B. Response to Central Walls Another set of observations is now advisable. In the first experiment the stimulus altered (presence or absence of a wall) was at the periphery of the field. What if you varied the test situation by placing walls in the center (CW condition) of the test arena? Would the mice still concentrate their activity around the edge of the field or would they be found near the centrally placed walls?

Use the same NW arena as in the previous experiment but place a square box 30 cm high over the four center squares (Fig. 13.3). Record data as before.

Total the number of observations for each animal in the 12 squares peripheral to the central wall as well as in the 28 squares around the floor perimeter. Compute means and standard errors for each sample.

Compare these sample means with the means obtained in the NW test condition for the same grid squares. Did the presence of the central wall increase the number of tallies in the squares immediately adjacent to it? Was there a corresponding reduction in the use of the 28 squares around the floor perimeter?

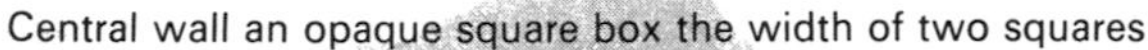

Figure 13-3. Central wall condition (CW).

QUESTIONS

1. Did the test mice show greater activity in areas immediately adjacent to walls? Interpret the differences in behavior observed in the W, NW, and CW conditions.

2. What selective factors could be responsible for the wall-seeking tendency of mice? What sensory modalities are probably involved? Why do you think so?

ADDITIONAL STUDIES

Mice are essentially nocturnal animals. How would their behavior in the test arena differ under different lighting conditions? How do other animals, such as roaches or lizards, react in the test arena?

REFERENCES

Fredericson, E., 1953. The wall-seeking tendency in three inbred mouse strains (*Mus musculus*). *J. Genet. Psychol.* 82:143–146.

Siegel, S., 1956. *Nonparametric statistics.* New York: McGraw-Hill.

Snedecor, G. W., and W. G. Cochran, 1967. *Statistical methods,* 6th ed. Ames: Iowa State University Press.

Thompson, W. R., 1953. The inheritance of behavior: Behavioral differences in fifteen mouse strains. *Can. J. Psychol.* 7:145–155.

LEE C. DRICKAMER
Williams College

Humidity Preferences of the Flour Beetle

Each species of animal and plant lives only where environmental conditions are within the tolerance limits permissible for that species. Most of these tolerances are functions of the inherited physiological and morphological systems of the different species, although certain tolerance limits may be altered by prior exposure to new or changing environmental conditions. Populations of a species that live under conditions that approach the tolerance limits for one or several of these environmental parameters are often fewer in number or the individuals are stunted in growth. For each species there are also optimal conditions, which lead to maximum production and population growth.

The process by which animal and plant species disperse and locate in diverse environments is called habitat selection. Habitat selection may be totally passive, as in many plant species and certain airborne insects, or it may be active, requiring movement by the animal in search of a suitable environment. This laboratory exercise demonstrates the selection process in one species of insect, the confused flour beetle (*Tribolium confusum*), for one environmental parameter, humidity. On the paired, segmented antennae of the adult *Tribolium* are a series of cells called hygroreceptors, which are sensitive to the amount of water in the air (Fig. 14.1; Willis and Roth, 1950). As the beetle moves about it turns its head from side to side, sampling the humidity of the environment. By this method the animal can turn toward the more favorable humidity and maintain itself in a tolerable environment. In this exercise you will test a series of adult beetles by giving them a choice between streams of air of different relative humidities.

METHODS

Subjects and Materials

The subjects to be used in the exercise are adult *Tribolium confusum* beetles. Several large colonies of the beetles should be started several months before the experiment to ensure adequate numbers of beetles. Generally several one- or two-gallon jars can serve as colony containers. Fill each jar three-quarters full with a mixture of 97% whole-wheat

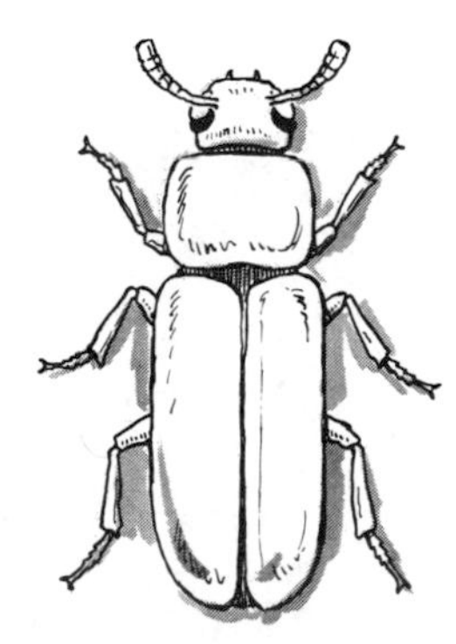

Figure 14-1. Adult confused flour beetle (*Tribolium confusum* DuVal).

flour and 3% brewer's yeast, introduce the starter population, and screw the cap on lightly. You may wish to add flour intermittently as the beetles consume the flour in the jar.

Each group of students will need 75–80 adult beetles for this exercise.

The equipment needed for the exercise includes:

Eight 500-ml Erlenmeyer flasks
Sulfuric acid (H_2SO_4)
Glass tubing
Rubber tubing
Two-hole rubber stoppers
Glass T-tube connectors
Masking tape
Small paintbrushes
Alcohol
Source of forced air

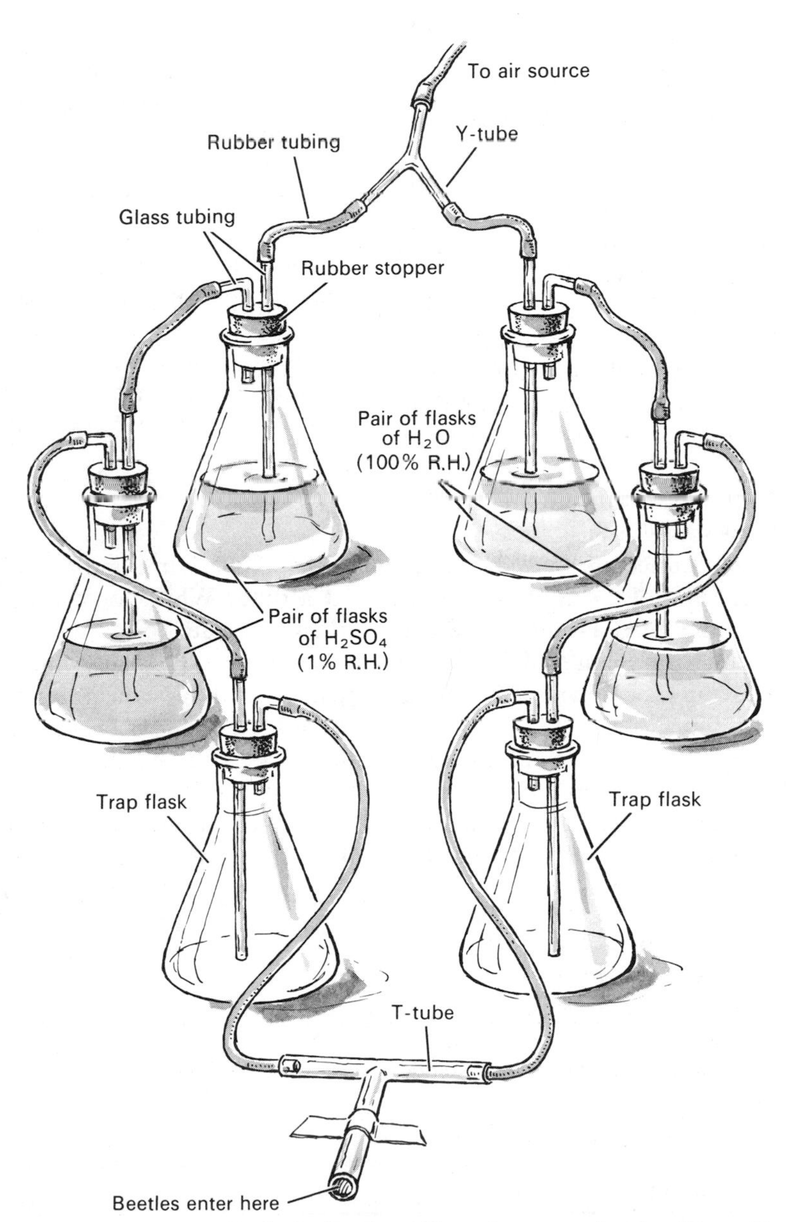

Figure 14-2. Schematic diagram of the apparatus for testing humidity preferences, showing the positioning and connections for each flask and the T-tube in which the beetle is tested. The set-up shown is for testing preferences for humidities of 1% and 100%.

Procedure

Place 250–300 ml of each of the following three liquids in each of two 500-ml Erlenmeyer flasks. Each liquid represents a different relative humidity (R.H.).

a. 1% R.H.: pure sulfuric acid, no water.
b. 50% R.H.: 43 grams of sulfuric acid to 57 grams of water. (*Caution:* Add acid to water slowly.)
c. 100% R.H.: pure water, no sulfuric acid.

Fit each flask with a 2-hole rubber stopper containing one short and one long glass tube, the latter extending to within one centimeter of the bottom of the flask (Fig. 14.2).

Using the rubber tubing and connectors, arrange your flasks in the parallel sequence shown in Figure 14.2 and connect the input hose to the forced-air source. Each pair of flasks on the same side should contain the same solution. By passing the air through both flasks you ensure that the moisture content of the air attains the prescribed level. Note that after passing through the two solutions the air goes through a trap flask to remove any impurities. This also acts as a safety precaution should you increase the air pressure too fast. By interchanging solutions you can pair all combinations of the three humidity levels.

Turn on the air lightly and count the air bubbles per unit time in the first solution flask on each side. Adjust your tubing and connectors so that the air flow is the same on both sides of the apparatus. Why is this necessary? Insert the glass T-tube at the end of the apparatus as shown in Figure 14.2, connect the air hoses to the ends and tape down the T-tube so that it lies flat on the table surface.

Once the apparatus is set up test 20–25 adult beetles with each of the three possible paired combinations of the three levels of relative humidity. Test the beetles one at a time and use a different beetle for each test.

To test a beetle use the paintbrush to remove the animal from the group and place the subject at the open entry port of the T-tube. Watch the beetle as it progresses down the tube to the junction. Note its head movements. At the junction the beetle will turn either left or right and move down the side-arm of the T. When the animal has passed a point 3 centimeters down one of the arms of the T, note which humidity level is selected, remove the T-tube and extract the beetle. Replace it with a clean dry tube and test the next subject.

T-tubes should be cleaned for each trial since the "trial" of odor left by one beetle might otherwise influence the choice made by the next test subject. Clean the T-tubes by washing them in water; then dip them briefly in alcohol for fast drying. If each student group has several T-tubes, one or two students can clean tubes while the others test the beetles.

Arrange your data as in Table 14.1, indicating which humidity levels were paired, the number of subjects tested, and the number and percentage of each test group selecting each humidity.

Chi-square tests (Siegle, 1956) may be performed on your data to determine whether the results are significantly different than chance (equal probability that a beetle would enter either arm of the T-tube).

QUESTIONS

1. Why is it important that you test only one beetle at a time? Why should a beetle be tested once and then discarded?
2. What can you conclude from your data and statistical analyses concerning the humidity preferences of adult *Tribolium* beetles? How might these results be related to the problem of habitat selection for this species?
3. What are some other behavioral methods used by different animal species in the process of selecting a preferred or optimum habitat?

Table 14-1. Humidity preferences of adult *Tribolium confusum* (flour beetles).

Percent R.H. pairs	*Number*	*Lower humidity*		*Higher humidity*	
		No.	*%*	*No.*	*%*
1 *vs* 100					
1 *vs* 50					
50 *vs* 100					

ADDITIONAL STUDIES

1. A second series of tests may be conducted using larvae of *Tribolium* in place of adult beetles. Or, half of the class may use adults and the remaining half larvae. The procedures are identical for both life stages. Do larvae exhibit humidity preferences different from those of adult beetles? What anatomical differences between larvae and adult beetles might influence their selection of levels of relative humidity?

2. The importance of hygroreceptors to the selection of different levels of relative humidity may be investigated by removing one or both antennae from adult beetles.

3. In addition, various aspects of orientation in the beetles may be explored. What physiological and behavioral processes are included in selecting a preferred level of relative humidity? The reference by Fraenkel and Gunn is an excellent starting place.

4. The experimental design and animal subjects used in this experiment can be used to test other environmental parameters such as light levels, surface textures, gravity, or temperature.

REFERENCES

Drickamer, L. C., 1971. The humidity preference of *Tribolium confusum* DuVal in wheat flour, sand and air. *Ohio J. Sci.* 71:149–158.

Fraenkel, G. S., and D. L. Gunn, 1961. *The orientation of animals.* New York: Dover.

Roth, L. M., and E. R. Willis, 1951. Hygroreceptors in adults of *Tribolium* (Coleoptera, Tenebrionidae). *J. Exp. Zool.* 116:527–570.

Siegel, S., 1956. *Nonparametric statistics.* New York: McGraw-Hill.

Willis, E. R., and L. M. Roth, 1950. Humidity reactions of *Tribolium castaneum* (Herbst). *J. Exp. Zool.* 115:561–587.

CARL J. BERG, JR.
City College of New York
and
The American Museum of Natural History

Orientation to Physical Conditions by Terrestrial Isopods

Sowbugs, pillbugs, and woodlice are terrestrial isopods (Phylum Arthropoda) that possess gills and are commonly found under leaf litter and debris. These animals orient to physical stimuli in their environment by a complex system of movements that are either directed with respect to the stimulus (taxes) or are undirected (kineses). Undirected responses are mediated by altering either the speed of locomotion (orthokinesis), the rate of change in direction (klinokinesis), or both (Fraenkel and Gunn, 1961). However, recent authors ask if klinokinetic responses may simply represent alternating directed responses (Stasko and Sullivan, 1971). The combined effects of many external environmental stimuli and many types of movements form the complex orientation shown by the animals we will study here.

This exercise is designed to show the effect of humidity and light intensity upon the locomotor patterns and distribution of isopods. You will study (1) differences in easily quantified behavior patterns exhibited under each set of environmental stimuli, and (2) the effects of the interaction of the stimuli. Because of the difficulties in detecting and controlling for stimuli eliciting directed responses, care should be taken in interpreting the results as examples of either taxes or kineses.

METHODS

Subjects and Materials

Any of the isopods of the suborder Oniscoidea are suitable (*Oniscus, Porcellio, Armadillidium*). A total of approximately fifty adult animals of the same species are needed for all four teams into which the class will be divided. The animals can be maintained in large stacking dishes or buckets with moist toweling or leaf debris. The containers must be kept covered to maintain high humidity. The animals should be acclimated to room temperature at least one day before testing.

Materials needed for this exercise include

- Eight 30-cm diameter stacking dishes
- Clear plastic or glass tops for dishes
- Four stopwatches
- A wax marking pencil
- Paper towels or filter paper
- Flat-black spray paint
- Vaseline or stopcock grease (for the edge of the dishes)
- Dry calcium sulfate or any other granular desiccant (thoroughly dried and kept at room temperature)
- Modeling clay
- Metal window screening

The stacking dishes, which will serve as experimental testing chambers, should be painted flat black on the inside. Obtain clear plastic or glass tops for the dishes and cut metal window screening out to fit inside the dishes. Filter paper or paper towels should be cut to fit under the screening, and modeling clay should be rolled into strips to seal the edges between the dish and the screening.

The eight test chambers are set up as follows:

2 humid (with wet paper towels on the floor).

2 dry (with desiccant on the floor).

2 half humid and half dry (with wet towels on one half and desiccant on the other, separated by a ridge of modeling clay as illustrated in Fig. 15.1).

1 dry with one half of the top painted black.

1 humid with one half of the top painted black.

The chambers must be set up at least two hours before the start of the laboratory period.

Procedure

The class should be divided into four teams of two students each, or multiples thereof. The first person in the team will run the experiments assigned to that team and the second member of the team will assist the first by timing the experiments and recording data. After each trial the roles may be reversed. Each student should perform the experiments twice, with a different animal each time. Data from the entire class will be compiled near the end of the laboratory period.

The isopods are best handled gently with your fingers or a spatula, not with forceps. Select active animals from the culture dishes. Never use an animal twice; discard the subjects at the completion of each trial. Be careful not to disturb the animals with shadows, loud noises, or vibrations of the table during the trials.

It is important to remember to keep the tops on the test chambers at all times in order to maintain the proper humidity.

A. Measurement of Behavior Patterns under Various Conditions Test conditions for the four teams:

Team A. Light–Dry

Team B. Light–Humid

Team C. Dark–Dry

Team D. Dark–Humid

The "Dark" tests should be run in a separate, completely darkened room with the aid of a low-wattage red light.

One animal is placed in the center of the chamber and the following data are collected for ten minutes:

Path of the animal. (You can record the path of movement on a piece of graph paper. A grid drawn on the glass top is helpful.) From your record you can then determine the rate of locomotion (distance covered per unit time) and the rate of change in direction (number of turns along any plane per unit time). For this latter measure, do not count slight body movements, only changes in direction of locomotion.

Total time spent moving. From these data compute the percentage of time spent moving by the following formula.

$$\frac{\text{Total number of seconds spent moving}}{\text{Total test period of 600 seconds}}$$

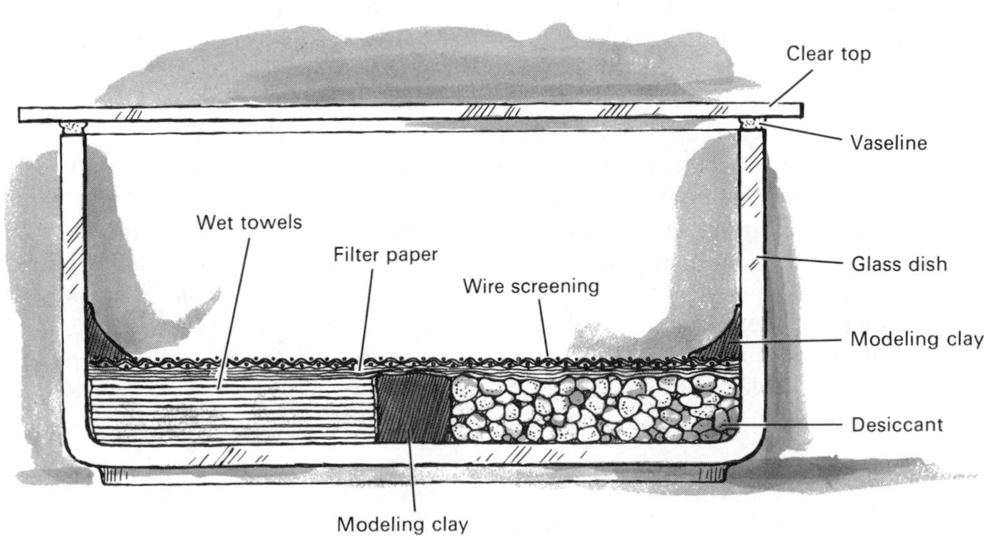

Figure 15-1. Test chamber with wet towels and desiccant to provide a humid : dry choice.

Total time within 1 cm of the edge of the clay. Calculate the percentage of time within 1 cm of the edge of the clay as you did for time spent moving.

Record on your data sheets the data for rate of locomotion, rate of turning, percentage of time moving, and percentage of time near the edge. Compute mean values for the compiled data and analyze for differences between adjacent blocks of data.

B. Preferences for Humidity and Light Test conditions for the same four teams:

Team A. Humid : Dry choice–Light

Team B. Humid : Dry choice–Dark

Team C. Light : Dark choice–Humid

Team D. Light : Dark choice–Dry

A line drawn with a wax pencil down the middle of the top helps to delineate the edge of the Humid : Dry gradient. The tops with one half painted black are used in the Light : Dark tests.

Testing procedures and data collection are identical to those in Part A, except that the data for each half of the test chamber are collected separately.

Compile separate tables of the data collected for each side of the test chambers. In addition, compute the total amount of time animals spent on each side of the test chambers and analyze for differences using chi-square techniques.

QUESTIONS

1. Under what conditions were the animals most active? In what different ways can you measure their activity?

2. Do these animals show differences in both the rate of locomotion and the rate of turning depending upon the environmental stimuli? How do the stimuli interact? How are the responses integrated?

3. How could undirected movements cause the animal to become confined to areas suitable for survival?

4. Devise experiments to determine if undirected (kinetic) responses are important in determining the distribution patterns of isopods. Try to explain how the animals become grouped in suitable areas, both in this experiment and in their natural habitat.

ADDITIONAL STUDIES

This exercise could be expanded in scope by using animals acclimated to a wide range of temperatures and relative humidity. Other stimuli might also be studied (e.g., gravity).

Do the animals studied show a tendency to group together? Is this caused by a positive thigmotactic (touch) response or a chemotactic response? Do they choose to be with members of their own species rather than other species or genera?

The sensory stimuli used in orientation can be studied by the removal or covering of the eyes and antennae. Advanced students may also try severing connections between the sensory apparatus and the central nervous system. How do these operations affect the orientation of the isopods and what insights do they give us concerning the mechanisms being used?

REFERENCES

Carthy, J. D., 1971. *An introduction to the behaviour of invertebrates.* New York: Hafner.

Cloudsley-Thompson, J. L., 1952. Studies in diurnal rhythms. II. Changes in the physiological responses of the woodlouse *Oniscus ascellus* to environmental stimuli. *J. Exp. Biol.* 29:295–303.

Den Boer, P. J., 1961. The ecological significance of activity patterns in the woodlouse, *Porcellio scaber. Arch. Neer. Zool.* 14:283–409.

Edney, E. B., 1954. Woodlice and the land habit. *Biol. Rev.* 29:185–219.

Evans, S. M., 1968. *Studies in invertebrate behavior.* London: Heinemann.

Fraenkel, G. S., and D. L. Gunn, 1961. *The orientation of animals.* New York: Dover.

Stasko, A. B., and C. M. Sullivan, 1971. Responses of planarians to light: an examination of klinokinesis. *Anim. Behav. Monogr.* 4:45–124.

Waloff, N., 1941. The mechanisms of humidity reactions of terrestrial arthropods. *J. Exp. Biol.* 18:115–135.

JOSEPH W. JENNINGS
University of Montana

16

Optical Orientation in the Blowfly Larva

Most insects respond to light stimulation: some positively, others negatively. In some insects light merely initiates movement; in others light orients movement as well.

Any directed movement or orientation in response to a light source is known as a taxis. A response by an organism possessing only a single light-intensity receptor (e.g., blowfly larva) has been referred to as a klinotaxis (Fraenkel and Gunn, 1961). In this exercise you will observe the response of blowfly (*Sarcophaga*) larvae to one and two sources of light arranged in different ways. After completing your observations you should consult the suggested references to help you identify and understand the behaviors observed.

METHODS

Subjects and Materials

Blowfly larvae may be obtained from many of the biological supply houses or from bait stores and pet shops. Consult Galtsoff, et al. (1959), if you wish to culture your own larvae.

The only materials needed are a large piece of black construction paper and two directional light sources (e.g., shielded lamps or flashlights).

Procedure

A. Response to Single Light Beam Place a single larva on a large piece of black construction paper. Place a single lamp several centimeters from the larva and at about a 30-degree angle with the horizontal (Fig. 16.1). Turn on the light and observe the larva's behavior. (If your larva is unresponsive, it may be ready to pupate and should be replaced.) Leave the light on until the larva has crawled off the paper. What direction did it take relative to the light source? Speculate on the mechanism by which the blowfly larva uses its single photoreceptor to orient to a light source.

Figure 16-1. Arrangement of a lamp to test the response of larva to a single light beam.

B. Response to Alternating Light Beams Arrange two lamps so that their beams intersect at right angles in the center of the paper (Fig. 16.2). Place the larva between the center of the paper and one of the lamps. Turn this light on and leave it on until the larva has moved several centimeters or to the middle of the paper in an oriented (nonrandom) fashion. Now turn the first light off and the second light on. Leave the second light on until the larva reaches an edge of the paper. Record its path. Take special note of any change in direction when the lights are switched.

C. Response to Intermittent Light Place the larva in the center of the paper with the light directly above. When the larva swings its anterior segments to one side, flash the light on and then off before the head is swung to the other side. Continue flashing the light on every time the larva swings its head in the one direction. By exposing the larva to light when its head is in different positions, you can alter the direction of movement.

D. Response to Two Light Beams Arrange the lamps so that their beams intersect about a quarter of the way along a diagonal line running between opposite corners of the black paper (Fig. 16.3). Place the larva near the corner closest to the intersection of the beams. Turn on both lamps simultaneously and closely observe the larva's behavior. When it has oriented and moved 10–12 cm, turn off the lamps, remove the larva, and change the arrangement of the lamps. Leave one lamp in its original location; move the other lamp farther away, but keep it directed so that the beams intersect at the same point as they did before. Put the larva back in its original position. Again, carefully note the larva's behavior when the lights are turned on.

Note any change from the previous path. Again, turn off the lamps and remove the larva. Now, take the lamp that is still closest to the diagonal, and move it as far as the lamp you reset previously. Again, place the larva near the closest corner, and closely observe the animal's path when the lights are turned on. Repeat each of the above light arrangements five or six times. Then draw the larva's "average" path for each of the three light arrangements. Compare these paths.

DISCUSSION

It is important to note that the larva cannot "see" the light source as an object at a distance from itself. It is capable only of detecting differences between the intensity of light stimulation falling on either side of its receptor. Thus, the world of light for the larva is one of gradients or degrees of intensity in otherwise diffuse illumination. The larva is, therefore, not goal oriented. The study by Loeb described by Fraenkel and Gunn (1961, p. 66) illustrates this situation.

QUESTIONS

1. Summarize the nature of the larva's orientation to light.

Figure 16-2. Arrangement of two lamps at right angles to measure the response of larva to alternating light beams.

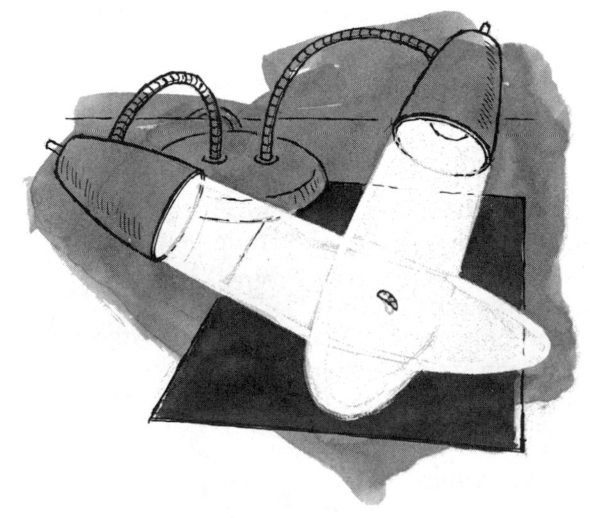

Figure 16-3. Arrangement to test the response of larva to intersecting light beams.

2. What is the name of this type of optical orientation? (See Fraenkel and Gunn.)

3. Of what use is this type of orientation to the larva under natural conditions?

4. What sensory and stimulus conditions must exist to allow a thermal and chemical klinotaxis?

5. How does the animal's orientation between two lamps demonstrate that its phototaxis is controlled by external stimulation?

6. Does the animal's tactic behavior indicate the existence of some form of retention?

7. What does the larva do when it bends its anterior segments that it could not do if its photoreceptor were on a stalk, like a ping-pong ball on a toothpick?

8. What change in the blowfly larva's sensory structure or function could eliminate the necessity for bending the anterior segments?

9. If the larva becomes unresponsive during this study, it may be habituating to the light. If a larva becomes unresponsive after a period in dim light, how could you tell whether it had become habituated or was beginning an irreversible process like pupation?

ADDITIONAL STUDIES

The larva's response to light is sufficiently reliable to use in assessing other response capacities of the organism. For instance, the tactic response can be used to indicate whether different light frequencies are equally bright to the animal. The question is, does the larva attempt to escape different light frequencies with equal vigor?

Red, blue, and green cellophane will be needed. Use other colors if they are also available. The cellophane is to be used for colored filters. Arrange your lights and larva as at the beginning of Part D: using three colors and two lamps, it is possible to arrange each of three different color pairs in two different left-right positions. Make sure that all the colored lights are of equal physical intensity, as determined by a light meter. If necessary, use layers of tissue paper to adjust intensity.

If each of a pair of colored lights is equally aversive, what path would you expect the larva to take in respect to the beams of the two intersecting lights? What can you tell about the relative aversiveness of the individual colors?

Almost any species of arthropod may be suitable for this type of orientation research. Try aquatic as well as terrestrial forms. Brine shrimp and *Daphnia* are readily available. How well suited are their forms of orientation to the environment in which they live?

REFERENCES

Carthy, J. D., 1958. *An introduction to the behavior of invertebrates.* London: George Allen and Unwin.

Denny, M. R., and S. C. Ratner, 1970. *Comparative psychology.* Homewood, Ill.: Dorsey Press. (See especially pp. 145–149.)

Dethier, V. G., 1963. *The physiology of insect senses.* New York: Wiley.

Fraenkel, G. S., and D. L. Gunn, 1961. *The orientation of animals.* New York: Dover. (See especially pp. 133–135.)

Galtsoff, P. S., F. E. Lutz, P. S. Welch, and J. G. Needham (Eds.), 1959. *Culture methods for invertebrate animals.* New York: Dover. (See especially pp. 414–427.)

Hinde, R. A., 1970. *Animal behavior,* 2d ed. New York: McGraw-Hill. (See especially pp. 151–161.)

Jander, R., 1963. Insect orientation. *Annu. Rev. Entomol.* 8:95–114.

Maier, R. A., and B. M. Maier, 1970. *Comparative animal behavior.* Belmont, Calif.: Brooks/Cole. (See especially pp. 79–83.)

Thorpe, W. H., 1963. *Learning and instinct in animals.* Cambridge, Mass.: Harvard University Press.

DALE L. CLAYTON
Walla Walla College

Rhythmic Behaviors

Our best evidence today suggests that environmental cycles are capable of synchronizing an organism's behavior, but that timing mechanisms within the organism enable it to anticipate cyclic changes of the environment. These timing mechanisms are called biological clocks.

The significance of this anticipation is apparent in the physiological preparation necessary for exploiting cyclic changes in the environment and for synchronizing the behaviors of individuals in a population. For example, the Palolo worm's spawning is largely limited to a brief period at dawn on a single day during the last quarter of the moon in November (Cloudsley-Thompson, 1961, pp. 87–88).

The fact that bees (von Frisch, 1964, pp. 345–346), frogs (Taylor and Ferguson, 1970), and many other organisms can orient by the sun and maintain a given compass direction all day implies that they can compensate for the sun's movement relative to the phase of the daily cycle. To do this they must have a clock (i.e., time sense). Can you think of other ways in which anticipating environmental changes or the synchronization of behaviors would be adaptive?

The objectives of this exercise are to demonstrate rhythmic drinking behavior in small rodents in both cyclic and constant environments, and to develop an awareness of rhythmic behaviors in general. The exercise described here requires a minimum of equipment and preparation. Concepts necessary for understanding rhythmic behaviors are described below and in Table 17.1 and Figure 17.1. A brief discussion of running-wheel activity in mice and emergence rhythms in fruit flies is given in Additional Studies.

Amplitude. The amplitude is the difference between the maximum and mean values of the phenomenon being measured. For example, if a mouse averages 5 wheel turns per minute (i.e., 7200 per day) and reaches a maximum rate of 100 wheel turns a minute, the cycle of running-wheel activity has an amplitude of 95 wheel turns per minute.

Period. The period is the duration of one complete cycle. The beginning and ending points of a cycle may be defined by discrete activities such as the onset and cessation of drinking bouts, running-wheel activity, and so on.

Phase. The term phase generally refers to an instantaneous point of reference in the cycle, but is sometimes used to describe longer intervals (i.e., light phase, dark phase, activity phase) in the cycle. A given phase is repeated once each cycle. The cycles

Table 17-1. Commonly studied cycle lengths.

Designation	*Length*
Infradian	Shorter than one day
Circadian	Approximately one day (20–28 hrs)
Ultradian	Longer than one day
Lunar	Monthly
Annual	Yearly

of two organisms are said to be in phase when specific features (e.g., the active or inactive phases) coincide. The onset of drinking bouts or of wheel-running may be used as phase reference points.

Synchronization (Entrainment). A biological rhythm (e.g., sleep-wake cycle) is said to be synchronized with or entrained to an external timer (e.g., light-dark cycle) if they both exhibit the same period and phase. For example, a mouse synchronized to a 12-hr light : 12-hr dark (12:12 LD) cycle in which the lights come on at 6 A.M. will be 180° (12 hours) out of phase with a 12:12 LD cycle in which the lights come on at 6 P.M. However, such a mouse will shift its phase to synchronize with the new light cycle, usually within 3 to 8 days.

An excellent and concise background for understanding rhythmic behaviors is provided by Marler and Hamilton (1966, Chaps. 2 and 4; see also Bunning, 1973).

METHODS

Subjects and Materials

Small rodents are excellent subjects for studying rhythmic behaviors; however, the following differences among them should be taken into account. Laboratory rats and mice exhibit less aversion to frequent observation than wild-caught animals and are preferred for studying short-term (infradian) cycles. Wild-caught rats and mice (particularly *Peromyscus*), hamsters, and squirrels tend to exhibit more distinct circadian patterns than laboratory rats and mice. Gerbils, kangaroo rats, and pocket mice have physiological mechanisms for conserving water, which limits the amount of drinking data that can be collected per unit time. Can you think of other parameters that should be considered?

Obtain a long piece of glass tubing (from 5 to 10 mm inside diameter and 50 cm to 1 m long). The tubing must have a large enough bore to allow bubbles to rise and not form an air lock. Make a gentle 45° bend near the bottom of the tube and fire-polish the end to narrow the opening slightly: this will prevent water from dripping out. Fill the glass tube with water and seal the top with the rubber bulb from a medicine dropper. You now have a water bottle that will enable you to measure the amount of water consumed with fair accuracy. Make a scale on paper and tape it behind the tube so that you can quickly determine the relative amount of water used. The volume may be calculated from the inside diameter of the tube but relative units are adequate. This very simple apparatus can be used to study rhythms of different period lengths (see Table 17.1).

Procedure

Studies of rhythmic behavior generally require more time than is normally allotted for a laboratory session. This becomes evident when you recognize that several cycles are required to establish the length of a period. If you and your classmates take turns observing during successive time blocks, longer observation is possible.

All animals should be maintained on an artificial 12:12 LD cycle for at least one week before testing to ensure entrainment (synchronization). An inexpensive timer can be used to turn lights on and off automatically. A dim light must be on during the entire dark (dim) phase to permit observation during testing. It is most convenient to let the dim light

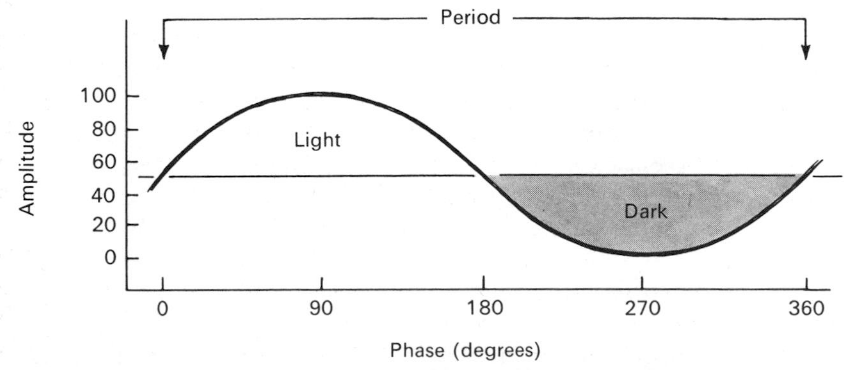

Figure 17-1. Three basic parameters of any cyclic or repeating function. A cycle has (by definition) a given *period* or time span during which it completes each repetition. The *amplitude* of the cycling variable is the sum of the distance it rises above and falls below the base line. *Phase* points are usually expressed in degrees, one period being represented by a full 360 degrees. In this diagram dawn occurs at 0° and dusk at 180°.

burn continuously and turn the brighter lights on and off. Cages should not be moved for observation but should remain where they were during the entrainment period.

Any disturbance (light, noise, etc.) can change the period, phase, or amplitude of the rhythm and make the data more difficult to interpret. Fortunately, rodents maintained under LD cycles are fairly resistant to mild disturbances: if maintained under constant conditions they are deprived of time cues and their rhythms are more labile.

Details for studying both infradian and circadian cycles are outlined below.

A. Infradian Cycles Mice should be observed a minimum of 4–5 hours. Longer observation is necessary for rats. At least three animals should be tested if each is observed for 5 hours or longer. More animals will be needed if observation sessions are shorter.

Nocturnal rodents are best observed during the dark (i.e., active) phase of their daily cycle. By reversing the light-dark cycle of your subjects your daytime will coincide with their dark phase, when they are most active and drink most. Prepare a data sheet for recording the following information.

1. Name of observer
2. Time and date when observations begin
3. Light : Dark conditions
4. Identification of data from individual animals
5. Onset time of individual drinking bouts
6. Amount consumed at each bout
7. Optional data

A drinking bout is defined here as the drinking phase of the cycle and is separated from the next bout by the nondrinking phase. Assign a minimal time value to the nondrinking phase. Nondrinking times shorter than the critical value are considered part of a continuous bout. Of course, observers allowing a shorter nondrinking phase may record more bouts than persons using a longer minimal phase. A minimal value of 15 seconds is useful but entirely arbitrary. What effect might a different value have on cycle length?

B. Circadian Cycles Circadian cycles of drinking can be demonstrated by recording the quantity of water consumed every 6 or 12 hours (more frequently if possible). Maintain a 12:12 LD cycle and record the water level at both the beginning and end of the light phase and other intermediate times that you decide on. Within a few days a pattern of drinking (water consumption) should become obvious. This pattern should suggest a definite phase relationship between the endogenous (behavioral-physiological) rhythm of drinking and the exogeneous (environmental) rhythm of the light cycle.

Is the light cycle necessary for maintenance of the rhythm of water consumption? To answer this question, place your subjects under constant dark (dim) illumination as soon as you have established the daily drinking rhythm under light-dark cycles. Unplug the cycling light only after the cycle has entered the dark phase, otherwise you may reset the phase of the drinking rhythm and make the results difficult to interpret. Record the following data under both cycling and constant dim light conditions.

1. Name of the observer
2. Time and date of observation
3. Light : Dark conditions
4. Identification of data with individual animals
5. Level of water in the tube (volume may be calculated later)
6. Optional data

Graph your data with time on the horizontal axis and volume drunk or bouts per unit time on the vertical axis. Data collected for infradian drinking rhythms should be plotted both in terms of bouts and volume. Do these two parameters exhibit different period lengths? Plot the data for each animal separately and look for individual differences in period, phase, and amplitude. This is particularly important for the infradian and free-running circadian data since a common synchronizing signal is not obvious.

DISCUSSION

In observing circadian rhythms we are able to eliminate the time cues (i.e., light cycle) and test for the existence of a free-running rhythm (i.e., a rhythm running free of external entrainment). This is not so easily done for infradian cycles, as the time cues are not apparent. We would expect the infradian and circadian drinking rhythms to be interrelated, but the details have not been elucidated. If you wish to test for a "biological clock" you must eliminate all environmental cycles impinging on the organism, otherwise the animal may be merely sensing and reflecting environmental changes.

Closely timed behavioral and physiological functions have been demonstrated in a wide array of plants and animals and yet no one has discovered the physical clocking mechanism or learned how it measures time. Some researchers do not use the term "biological clock" because it implies a discrete mechanism which, at this time, has not been identified, but the concept of a "clock" as a discrete entity has stimulated a great deal of research which, in turn, has advanced our understanding of rhythmic behaviors. Although the weight of present evidence favors the interpretation of an endogenous timing mechanism (i.e., a timer within the organism), we are tempted to ask what questions have gone unasked because the majority of researchers disregard the possibility of an exogenous (i.e., external) timer. Frank Brown (1972) and his students have been the major proponents of the exogenous hypothesis. Advocates of the endogenous hypothesis are too numerous to mention but certainly Pittendrigh (1956) has been a leading proponent. Whereas supporters of the endogenous hypothesis are impeded by the lack of evidence for a demonstrable "clock," the exogenous hypothesis rests entirely upon "subtle geophysical variables" which are also unidentified. In spite of the fact that the "clock" is an enigma, a great deal is known about its function. The gaps in our understanding provide interesting and challenging areas for investigation.

QUESTIONS

1. What rhythmic behaviors do you exhibit? What rhythmic behaviors have you observed in other animals? Are these rhythms adaptive? What additional experiments do they suggest?
2. What is the amplitude of the cycle in Figure 17.1 (mean = 50; maximum = 100; minimum = 0)? What would be the result of assigning a minimum time value to the nondrinking phase of the infradian cycle that exceeded the period of the cycle?
3. Do you see individual variability in the data of the different subjects? In what parameter (period, phase, or amplitude) do you see the most variability? What would this suggest in terms of natural selection on these parameters of biological rhythms?

ADDITIONAL STUDIES

Running-wheel activity in rodents. If you have access to an event recorder, running-wheel activity can be used to measure the circadian activity rhythms of many rodent species. The additional time you will spend in setting up the more elaborate apparatus you will save by automation. The considerations in the foregoing text on drinking rhythms are applicable and an extended discussion here is not necessary.

An adequate wheel can be constructed by making a wire mesh cylinder, 15–30 cm in diameter and 10–20 cm long, depending on the size of your animal. Obtain two pieces of wood as long as the diameter of the wheel and wide enough to allow the screen to be tacked securely to the ends. Drill a hole in the middle of each stick. Each hole should be slightly larger than the heavy wire or rod that will form the axle of the wheel. Support this axle at each end. A magnetic reed switch (see your local electronics dealer) is the simplest apparatus for detecting movement of the wheel. A magnet that adheres to the outside of the wheel will close a reed switch mounted above the wheel each time it passes. Power can be supplied to the event recorder through this reed switch and each turn of the wheel will be recorded. Mechanically operated switches and other devices are also used (Cloudsley-Thompson, 1961, Chap. 1).

The most commonly used phase reference for wheel-running is onset time (i.e., first turn after the inactive phase). Other parameters such as end of running, total turns each hour, and so on, are also of interest. (See DeCoursey, 1960.)

Eclosion time in Drosophila. A discussion of rhythmic behaviors would hardly be complete without mention of this frequently studied phenomenon. *Drosophila* sp. (fruit flies) can be obtained from biological supply houses, or almost any garbage can. Culture media are available also from supply houses. At 20°C and 25°C fruit flies develop from egg to adult in about 15 to 10 days respectively. As the flies begin emerging from the pupae cases the bottles containing the cultures may be cleared of adults periodically by removing the stoppers and inverting the bottles over empty collecting bottles.

Few flies escape if collections are made in the following manner. Remove the stopper from the culture bottle while striking the latter lightly on a book or table top. Invert an empty bottle quickly over the culture bottle and then turn them over so that the empty bottle is on the bottom. Bang the bottles up and down and then plug the bottom bottle quickly. (An alternative method for transferring flies is described in Exercise 20.) Record the number of flies that emerged each time and treat the data as suggested in the main exercise. Adults may be discarded or maintained as breeding stock.

Care must be taken not to introduce additional light cues; some of the collections must be made in darkness or in dim light. Collections every 4–6 hours (or more frequently) are recommended. (See Pittendrigh, 1958.)

In these additional studies, as in the study of drinking behavior, innate rhythms can be demonstrated only if data are collected under constant conditions.

REFERENCES

Aschoff, J., 1960. Exogenous and endogenous components in circadian rhythms. *Cold Spring Harbor Symp. Quant. Biol.* 25:11–28.

Brown, F. A., 1972. The "clocks" timing biological rhythms. *Amer. Sci.* 60:756–766.

Bunning, E., 1973. *The physiological clock.* New York: Springer-Verlag.

Cloudsley-Thompson, J. L., 1961. *Rhythmic activity in animal physiology and behavior.* New York: Academic Press.

DeCoursey, P., 1960. Phase control of activity in a rodent. *Cold Spring Harbor Symp. Quant. Biol.* 25:49–55.

Frisch, K. von, 1964. *Biology.* New York: Harper and Row.

Marler, P., and W. J. Hamilton, 1966. *Mechanisms of animal behavior.* New York: Wiley.

Pittendrigh, C. S., 1958. Perspectives in the study of biological clocks. *In* A. Buzzati-Traverso (Ed.), *Perspectives in marine biology,* pp. 239–268. Berkeley: University of California Press.

Taylor, D. H., and D. E. Ferguson, 1970. Extraoptic celestial orientation in the southern cricket frog *Acris gryllus. Science* 168:390–392.

EDWARD O. PRICE
State University of New York, Syracuse
WILLIAM H. CALHOUN
University of Tennessee
CHARLES H. SOUTHWICK
The Johns Hopkins University
THOMAS E. McGILL
Williams College

18

Behavior Genetics of Inbred Mice

Inbred strains of mice (*Mus musculus*) offer an opportunity to study the interaction of heredity and environment in the development of behavior. Inbred lines and their F_1 crosses are isogenic: that is, all individuals within each stock are genetically similar. Inbred strains are theoretically homozygous at all loci, whereas their F_1 hybrids are uniformly heterozygous for all loci in which the parent strains differ. Since genetic variation within isogenic lines is minimal (theoretically zero), any phenotypic variability observed between individuals within these lines is considered primarily due to environmental influence. Conversely, differences between individuals of different isogenic lines are considered genetic, assuming, of course, that they have had the same environmental experience. To go one step further, if isogenic lines are subdivided and subpopulations are given different environmental experiences, one can study the different ways that various genotypes interact with the environment.

Genes do not affect behavior directly. Rather their effect is mediated through enzyme activity, hormone levels, tissue sensitivity, membrane permeability, and other functions and structures (Thiessen, 1972, p. 24). Consequently, most behavioral traits are influenced by many genes (are polygenic) and show various degrees of expression when scored on a continuous quantitative scale. When two inbred strains of mice are cross-mated the behavior of the hybrids may be related to that of the parent strains in one of three ways. First, the hybrids may be intermediate to the parent strains. With intermediate inheritance the genes from each parental stock may be thought of as opposing (balancing) each other. Second, the hybrids may resemble one of the parent stocks but not the other. Under such conditions of dominance the genes of one parent mask or block the expression of the genetic complement from the second parent. A third possible outcome is heterosis, or "hybrid vigor," in which the hybrid scores are more extreme than those of either parent. Heterosis is thought to occur only for traits that are subject to inbreeding depression (an overall reduction in "fitness" resulting from the expression of deleterious recessive genes—caused by inbreeding). Heterosis reflects "directional dominance" whereby the genes from the two parent lines, instead of opposing one another, influence a given character in the same direction.

This exercise describes three experiments that can be conducted with inbred mouse strains to demonstrate the behavioral differences between mice of different genotypes and to show how the genotype can determine the relative impact of early social experience on adult social behavior.

OPTION I. OPEN-FIELD BEHAVIOR

The objective of this experiment is to assess the effect of genotype on gross activity, climbing, and defecation rate in an unfamiliar test chamber (open-field).

METHODS

Subjects and Materials

Obtain a minimum of 6 mice of the same sex from each of two inbred mouse strains and their respective hybrid (C57BL/6, A/J and their hybrid, $B6AF_1$, are recommended). It is assumed that all mice used have had the same rearing history. If all mice are placed in individual cages just before testing, individuals need not be marked for identification. Subjects should be 40–100 days old when tested. Handling should be standardized (e.g., grasping tail by hand or with large rubber-tipped forceps).

Each group of students will need a stopwatch, an observation chamber (e.g., 5-gallon aquarium), and a piece of relatively stiff hardware cloth (wire mesh) that extends diagonally from the bottom of one side of the chamber to the top of the other side as shown in Fig. 18.1. The floor of the chamber should be marked off into a grid of 5- to 8-cm square sectors.

Procedure

Each student team will be assigned one or more mice of each genotype. Place a single mouse on the floor of the observation chamber and then insert the wire mesh. Each mouse is tested for 15 minutes in the chamber with the following data being recorded: (1) time spent climbing on the wire in seconds (all 4 feet must be on wire), (2) number of activity units (floor sections) traversed (this score should be adjusted for total time spent on the floor of the chamber: total time of test minus time on wire divided into total number of activity units traversed) and (3) number of fecal boluses deposited. If dim lighting is used during testing the mice will be relatively insensitive to the observer's presence. Even so, avoid talking and sudden movements. The observation chamber should preferably be cleaned after each test.

Mice should be tested only once per day. If possible, test each mouse on three different days and compute its mean score for each variable. The data for all mice should then be pooled and analyzed by analysis of variance techniques or by simple *t* tests.

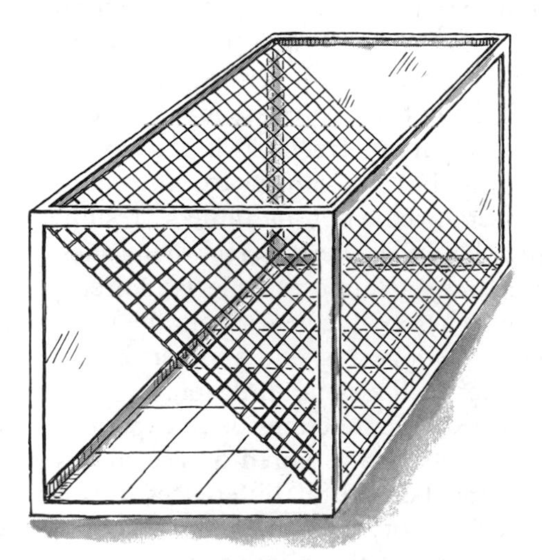

Figure 18-1. A suggested open-field apparatus with grid floor and wire-mesh climbing surface.

DISCUSSION

The open-field test is one of the most frequently used techniques for measuring exploratory activity in a novel environment. Defecation rate is sometimes used as an index of emotionality. (This latter term should be used with caution since it lacks the critical objectivity discussed in Exercise 1 of this manual). For information on the validity of open-field activity and defecation as measures of emotionality, see Archer (1973).

Climbing is normally not measured in open-field experimentation but is included here as an additional variable by which to assess genotypic effects on behavior.

QUESTIONS

1. What were the effects of genotype on the variables measured?
2. Which of the three types of inheritance did the F_1 hybrids exhibit for each of the variables measured? Can you explain why the type of inheritance differed for different variables?
3. Would you expect the parent strains or the F_1 hybrids to show the greatest variability in behavioral scores? Why? Do your data support your hypothesis?
4. Did you notice any differences in the ease of handling between the various genotypes used? How do these differences correlate with open-field activity and defecation?

OPTION II. AGONISTIC BEHAVIOR

The objective of this experiment is to assess the effect of genotype on the frequency of various agonistic behavioral patterns and to study the interaction between genotype and early social experience.

METHODS

Subjects and Materials

Eighteen male mice of each of two genotypes are needed for this experiment. Rear nine mice of each genotype in social isolation (i.e., isolation of individuals in single cages) from weaning (20–25 days) until sexual maturity (40–50 days) or longer, and nine in groups of three. Mark all individuals for positive identification (e.g., temporary identification—felt tip pen: permanent—ear notch, toe clip).

Each group of students will need a stopwatch and a dimly lighted observation chamber (e.g., 5-gallon aquarium; clear plastic cage or cylinder, as in Fig. 18.2).

Procedure

Weigh three mice (social strangers) of like genotype and treatment and place them in the observation chamber. Note their initial reactions toward each other. During the ensuing 30-minute period record the following data: (1) latency for the first aggressive interaction (chase, attack, or fight), (2) number of chases, (3) number of attacks (one mouse showing active aggression, usually with bites or physical contact, without the other reciprocating), and (4) number of fights (physical assault by two or more individuals usually including wrestling). The number of nose-to-nose and nose-to-genital contacts may also be recorded, as may the frequency of various postural displays (Fig. 18.3). Then return the mice to their respective home cages. See Clark and Schein (1966) for further discussion of the variables employed.

Since it is desirable for you to observe at least one triad of each of the four genotype–treatment combinations, mice may be regrouped (three social strangers) during the same laboratory period, providing a minimum of 30 minutes has elapsed between tests. However, testing each individual only once per day is preferable.

Pool the data for each genotype–treatment combination and analyze by appropriate statistical procedures (e.g., analysis of variance).

DISCUSSION

Certain kinds of behavioral data (e.g., open-field activity, climbing, etc.) are discrete, with minimal

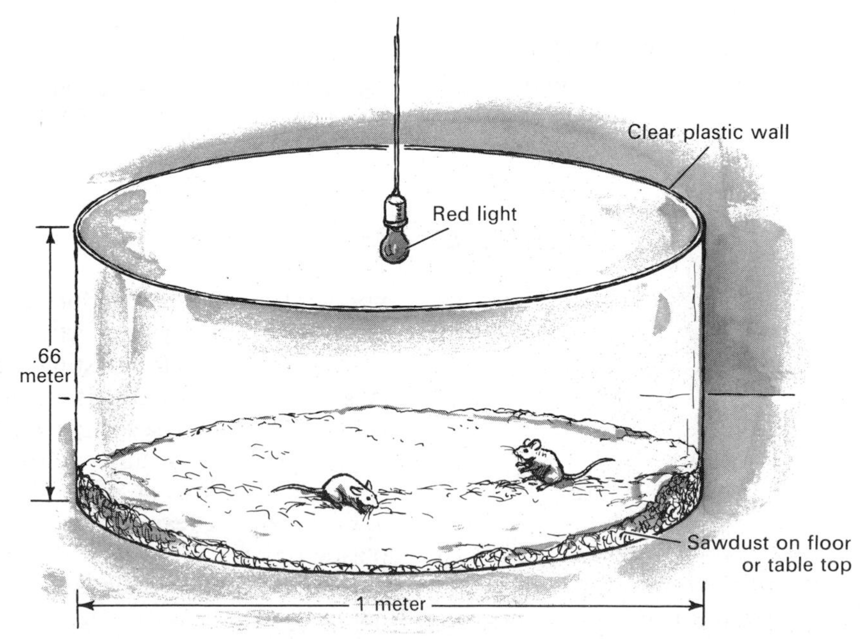

Figure 18-2. Circular arena for observing social interactions of mice.

opportunity for observational error. Other variables (e.g., certain social behaviors such as chasing, attack, etc.) are often not expressed in a clear, concise manner and may be subject to observer interpretation. Such experimental error is difficult to avoid, particularly when different individuals are collecting data in the course of a single experiment. To what extent do you think the variability of the data collected by you and your fellow students in this experiment is due to poor interobserver reliability?

QUESTIONS

1. In what ways did genotype and treatment influence the social behavior of your mice?
2. To what extent was the effect of early social experience dependent on the genotype of mice employed?
3. What were the initial reactions of the mice when grouped together?
4. Did all individuals engage in fighting? Were some mice submissive from the start? To what extent was body weight correlated with dominance in aggressive interactions?
5. Under what conditions were vocalizations emitted?

ADDITIONAL STUDIES

One obvious extension of the present study would be to observe social behavior in cross-strain interactions. Is one genotype generally dominant over other genotypes? How is cross-strain dominance influenced by early social experience?

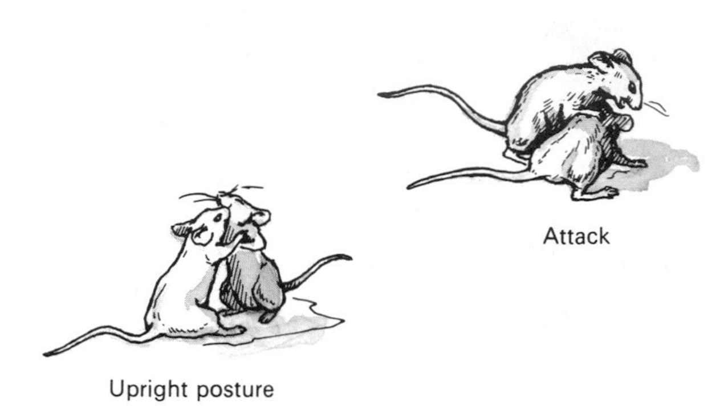

Figure 18-3. Some social postures of the mouse, *Mus musculus.* (Modified from Grant and Mackintosh, 1963.)

OPTION III. SEXUAL BEHAVIOR

The objective of the following experiment is to assess the effect of genotype on the sexual behavior of *Mus musculus.*

METHODS

Subjects and Materials

Ten male mice of each of two inbred lines (C57BL/6 and DBA/2 are recommended) and their respective F_1 hybrid (B6D2F_1) are needed, as well as about 24 female mice (BALB/c are recommended). The experiment should be conducted when the mice are from 50 to 180 days old. All males should be caged singly two days before testing. Females may remain in groups of four to six. If possible, place the entire colony on a reversed light–dark cycle about two weeks prior to testing. Make your observations during the dark (dim light) phase.

Each student team will need a stopwatch, a clear plastic rodent cage, a 1-cc tuberculin syringe with a 2.54-cm 25-gauge needle, and a blunt probe. A small quantity of estrogen is needed; any one of several different varieties will probably induce estrus in the females. However, "Progynon Benzoate," a brand of estradiol benzoate, available from Schering in 1 mg/cc concentration, is recommended.

Procedure

A. Preliminary Preparations Females are injected with .01–.02 cc of the 1 mg/cc concentration of estradiol benzoate 24–36 hours before testing. Injections should be made intramuscularly into the flanks of the females. This may be accomplished in either of two ways. The female may be held by the tail and placed on a rough surface that offers good footing to the animal. When her tail is pulled gently, the female will tend to pull away from the hand holding her. As the female pulls away the needle is inserted into the fleshy part of the flank and the hormone injected.

Occasionally a female will turn around and bite the needle during the injection. If this proves troublesome, a second technique may be used (Fig. 18.4). Thumbtack two pieces of stiff plastic or cardboard side by side to the edge of a table. These should extend about 15 cm above the top of the table, and should be about 6–7 mm apart from the top downwards to a point about 2 cm from the table top. At this point, each piece of plastic or cardboard should be cut back another 6 7 mm to create a 2 cm opening. A female may then be picked up by the tail and lowered between the two pieces of cardboard so that her body is on the table side of the cardboard, while her tail in the experimenter's hand is on the other side. After she has been lowered to the table, she may be pulled by the tail until her flank is accessible through the opening. The injection is then easily given in the flank and the animal is unable to bite at the needle.

If a male is given sexual experience before the actual experiment, he will probably mate more readily during the experiments. Females will exhibit slightly better estrus if they have received weekly injections of the hormone for two or three weeks. Even so, if time permits, it is advisable to run preliminary tests on the animals. To do this, place an injected female into the male's home cage in the morning. Remove the female in the late afternoon or evening. When the female is removed, examine her for the presence of a copulatory or vaginal plug. (Part of the semen of a male mouse forms a hard, rubbery plug in the vagina of the female.) You can detect a plug by inserting a blunt probe into the vagina. If a plug is present, it may be visible as a hard white ball in the vagina. Even when not visible, the plug may be detected by the resistance that it offers to the probe. In learning this technique, it is helpful to "probe" females that have not been with males in order to become familiar with the vagina in its unplugged state.

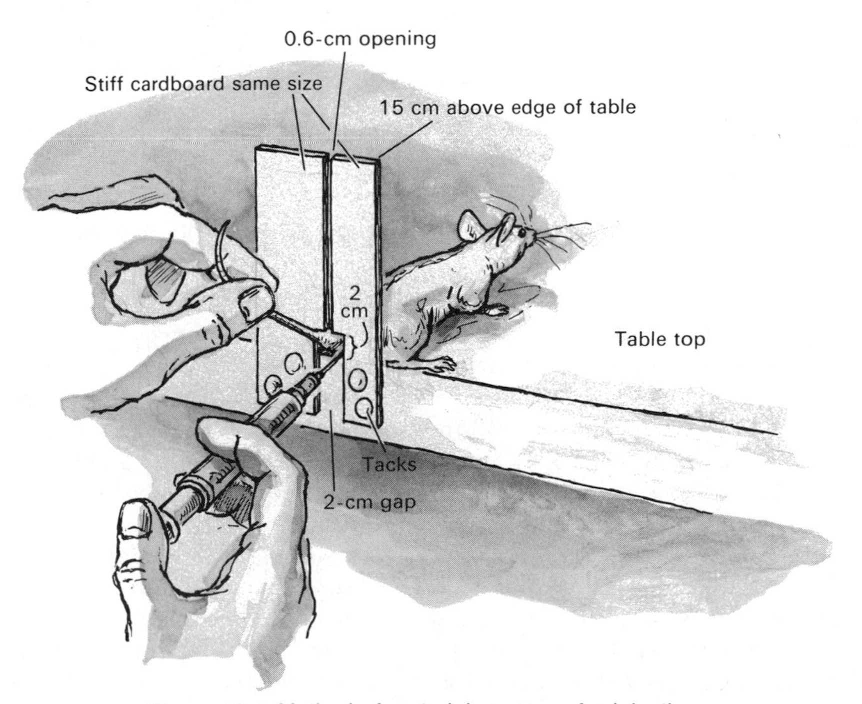

Figure 18-4. Method of restraining mouse for injection.

B. Testing On the day of the experiment, place the cages of the males on tables in the laboratory. Illumination should be as dim as possible. Each team should have one male of each genotype. None of these males should have ejaculated for at least one week before the test. The males should be allowed 15–20 minutes to adapt to the new situation and to the light.

After the adaptation period, select a male at random and introduce a receptive estrous female into his cage. Start the stopwatch.

Record the elapsed time at the beginning of the first mount without stopping the watch. Similarly, record the time of the first thrust-of-intromission without stopping the watch.

An unsuccessful attempt by the male to penetrate the female should be considered a mount. A successful attempt is an intromission. When the male mounts, he palpates the female's sides with his forepaws and executes a series of very rapid, probing, pelvic thrusts. If he is successful in gaining intromission the speed of thrusting is reduced to about two thrusts per second. These "thrusts-of-intromission" are easily counted and easily distinguished from the probing thrusts of a mount.

Once a male begins to mate, he will probably continue until ejaculation occurs. Occasionally, however, a male will cease his attempts to copulate. If thirty minutes pass without an intromission, the copulation may be called "incomplete" and the female removed.

As a male mouse approaches ejaculation, the speed of thrusting is increased; then the male quivers strongly, clutches the female with all four limbs, and falls to his side. At the first shudder, note the time on the stopwatch. When the male and female separate following the ejaculation, stop the stopwatch.

Note the number of thrusts making up each intromission during the copulation. Record the number of mounts, or unsuccessful attempts to penetrate the female.

If a male does not initiate sexual behavior within a 15-minute period with the first female, remove that female and immediately offer a second female to the male. If the male "passes up" the second female, try a third and final female.

For each male, record the following measures:

1. The number of females "passed up."
2. Mount latency: the number of seconds from the introduction of the female until the first mount by the male.
3. Intromission latency: the number of seconds from the introduction of the female until the male first gains intromission.
4. The number of intromissions preceding ejaculation.
5. Ejaculation latency: the number of seconds from the first intromission to the beginning of ejaculation.
6. The number of mounts during the copulation. Also of interest is the number of copulators in each strain.

Group these data on the board and apply appropriate statistical tests in a search for strain differences. After the male ejaculates, remove the female and look for the presence of a vaginal plug.

If time permits, place a fresh estrous female in the cage with the male one hour after ejaculation and repeat the test. Record the same measures as before. If the male copulates for a second time, compare his scores on the various measures of sexual behavior with those recorded during his initial copulatory behavior.

QUESTIONS

1. Which of the genotypes had the greatest percentage of copulators? Of those mice that copulated, which genotype exhibited the most mounts and most intromissions prior to ejaculation?

2. How did the hybrids compare with the parent strains? What inferences can you make regarding the mode of inheritance of the variables measured?

3. On the basis of your results how well do you think the various genotypes would compete with one another in mating with an estrous female?

ADDITIONAL STUDIES

One obvious extension of the present exercise would be to place simultaneously one male of each genotype together with a single estrous female to determine competitive advantages in mounting, in gaining intromissions, and in ejaculation. Another study would be to pair noninjected females with inbred males possessing different coat-color markers. Thus, strain differences in mating success could be assessed by the pelage coloration of the offspring. For example, an A/J female mated with both a C57BL/6J and an A/J male would produce black offspring from the former male and white offspring from the latter male.

REFERENCES

Archer, J., 1973. Tests for emotionality in rats and mice: A review. *Anim. Behav.* 21:205–235.

Clark, L. H., and M. W. Schein, 1966. Activities associated with conflict behavior in mice. *Anim. Behav.* 14:44–49.

Grant, E. C., and J. H. Mackintosh, 1963. A comparison of the social postures of some common laboratory rodents. *Behaviour* 21:246–259.

Green, E. L. (Ed.), 1966. *Biology of the laboratory mouse.* New York: McGraw-Hill. (See especially Chap. 33.)

Hirsch, J. (Ed.), 1967. *Behavior–genetic analysis.* New York: McGraw-Hill. (See especially Chaps. 13, 15, and 16.)

Lindzey, G., and D. D. Thiessen (Eds.), 1970. *Contributions to behavior–genetic analysis: The mouse as a prototype.* New York: Appleton-Century-Crofts. (See especially Chaps. 1, 2, and 3.)

Manosevitz, M., G. Lindzey, and D. D. Thiessen (Eds.), 1969. *Behavior genetics: Method and research.* New York: Appleton-Century-Crofts. (Contains many chapters relevant to this exercise.)

McGill, T. E., 1970. Genetic analysis of male sexual behavior. *In* Lindzey, G., and D. D. Thiessen (Eds.), *Contributions to behavior–genetic analysis: The mouse as a prototype,* Chap. 3. New York: Appleton-Century-Crofts.

Southwick, C. H. (Ed.), 1970. *Animal aggression: Selected readings.* New York: Van Nostrand Reinhold. (See especially Chaps. 7, 14, and 15.)

Thiessen, D. D., 1972. *Gene organization and behavior.* New York: Random House. (Excellent reading for the interested student.)

Thompson, W. R., 1953. The inheritance of behavior: Behavioral differences in fifteen mouse strains. *Can. J. Psychol.* 7:145–155.

Vale, J. R., C. A. Vale, and J. P. Harley, 1971. Interaction of genotype and population number with regard to aggressive behavior, social grooming, and adrenal and gonadal weight in male mice. *Commun. Behav. Biol.* 6:209–221.

DAVID P. BARASH
University of Washington

Reaction Chains in Insect Courtship: *Nasonia vitripennis**

The behavior of many animals can be viewed as the result of relatively automatic, unvarying, and genetically determined responses to particular stimuli. The "classical ethologists" Lorenz (1952) and Tinbergen (1951) have accordingly developed a conceptual framework with a terminology of "releasers," "IRM's," "action specific energy," "displacement activities," and so on. Although most ethologists have become increasingly dissatisfied with this system, it does provide a coherent explanation for much behavior. For a good general treatment of the classical model by a tenacious adherent, see Eibl-Eibesfeldt (1970). This classical model can generate complex and delicately coordinated behaviors between two animals, if each responds automatically to the behavior of the other and that response in turn "releases" the next step in the sequence, and so on. By careful observation of the temporal sequence of behaviors, a "reaction-chain" diagram can often be constructed:

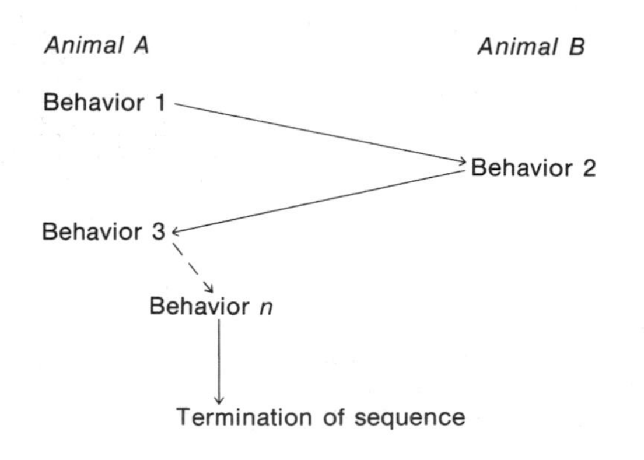

Such analysis requires both identification of each behavior unit (classically, the "fixed action patterns") and close attention to "what follows what." Of course, the final diagram is often more complex than the schematic one shown above. In particular, an animal may behave in different ways, depending upon its internal state ("motivation") and its prior experience. This requires the addition of extra arrows, and often requires recycling the sequence to an earlier stage or even terminating the reaction chain before reaching behavior *n*.

This laboratory exercise provides a demonstration of reaction-chaining in the courtship behavior of a harmless *Drosophila*-size wasp, *Nasonia vitripennis,* which parasitizes various fly pupae.

METHODS

Subjects and Materials

You will be provided with two stoppered vials, each containing virgin adult *Nasonia*, at least 6 males in one vial and an equal number of females in the other. A binocular dissecting microscope, or a tripod-mounted or even hand-held magnifier may be used for observation.

Procedure

One male and one female are introduced into a small Petri dish (the smaller the better, to facilitate encounters) with a tight-fitting lid. Because the females

*For many years this species was referred to as *Mormoniella vitripennis*. (*Mormoniella* is a synonym of *Nasonia.*)

can fly, whereas the males are flightless, it is best to introduce the males first, to minimize escapes. However, both males and females may be handled like *Drosophila* (see Exercises 20 and 21), tapped to the bottom of the vial and then shaken out as desired.

Within a few minutes (sometimes seconds) after introduction of both sexes, courtship will commence. Concentrate your initial attention upon the gross aspects of behavior, identifying the basic courtship and copulatory postures (Figs. 19.1,a and b). Then, squash the animals with a cotton ball and introduce another pair, looking now for the finer details of the courtship behavior. Exactly what does the male do? The female? (Fig. 19.2). If a post-copulatory courtship occurs (as it frequently does) is it followed by another copulation? If not, is this because the male attempts it but is actively rebuffed by the female, or because the male does not attempt a second copulation? If the latter, is he responding to a particular signal (or lack of signal) from the female? To answer this, compare the initial behavior of a virgin female with that of a nonvirgin. (Virgin *Nasonia* females are almost always sexually receptive, whereas nonvirgins are generally unreceptive.) When you have answered these questions, you should be able to construct a reaction chain for *Nasonia* courtship.

By means of a simple experiment, you can investigate the role of specific stimulus input in coordinating male–female behavior and maintaining the reaction chain. The antennae are clearly involved in courtship, and they may be removed by enmeshing the animal in cotton and carefully twisting them off with fine forceps. This is a difficult procedure to perform correctly–that is, without injuring the animals. Therefore it will probably be best to concentrate your efforts on two or three good amputations, relying on pooled class data for a statistically reliable sample. How does the courtship of amputated males differ from that of intact animals? How does amputation of female antennae affect courtship of intact

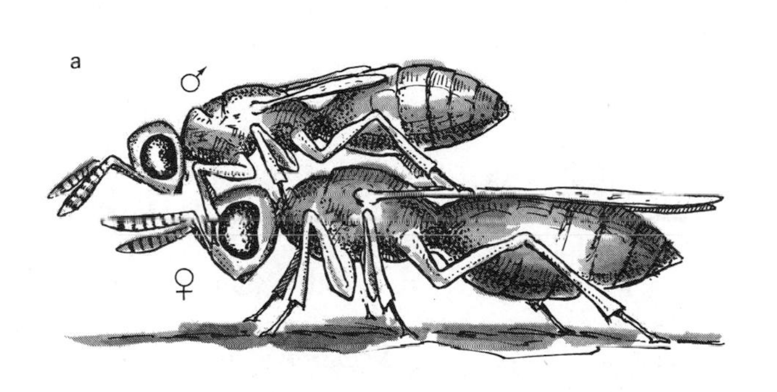

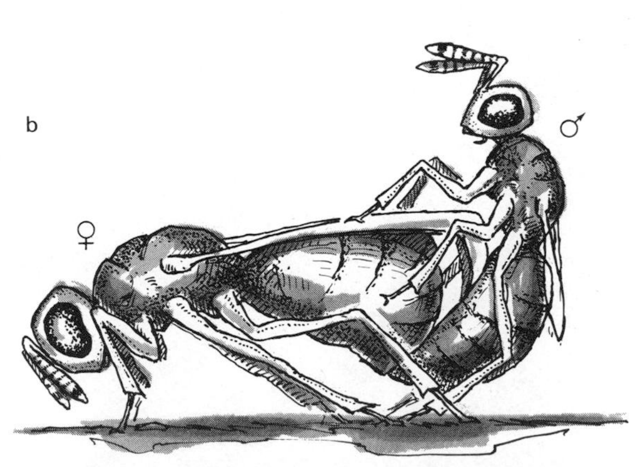

Figure 19-1. (*a*) Male above female in courtship position; (*b*) Male behind female in copulatory posture. Note the female's lowered antennae and elevated abdomen and the contact made at the female's genital pore. (After Barrass, 1960.)

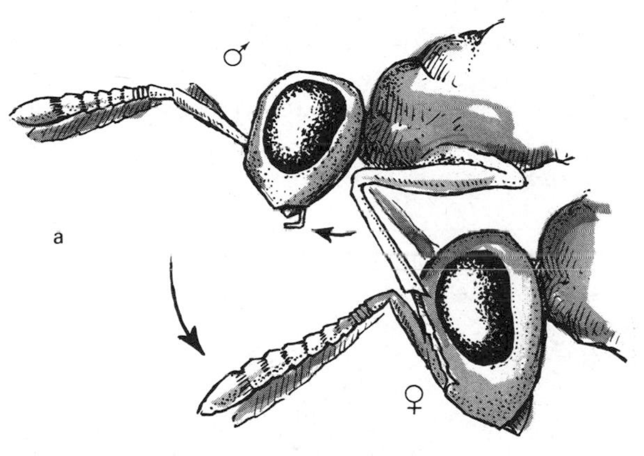

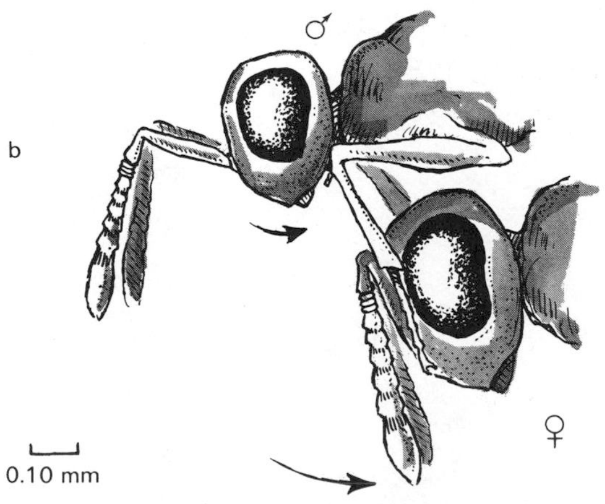

Figure 19-2. Lowering of female's antennae in response to courting male. Note extension and retraction of male mouth parts. (After Barrass, 1960.)

males? How does amputation of the antennae of virgin females affect their receptivity? Nonvirgin females? (In interpreting your results, remember that the antennae serve both to *receive* and to *transmit* relevant stimuli.)

QUESTIONS

1. Why are reaction chains uniquely suited to descriptions of courtship behavior? A slightly different, but also significant question: Why does animal courtship so often utilize behaviors that are describable in terms of reaction chaining?
2. What is a possible evolutionary advantage of reaction chains? (Consider the significance of isolating mechanisms.) A potential disadvantage?
3. What relationship exists between a species' capacity for modifying its behavior and its reliance upon reaction chains?
4. Can you identify reaction chains in courtship of fish? amphibians? birds? mammals? human beings?
5. Does previous exposure to parts of the reaction chain reduce the responsiveness of *Nasonia* males to subsequent exposures?

ADDITIONAL STUDIES

Your work so far should suggest many more questions. Is there any change in male responsiveness following successive copulations? Does age of the male or the female influence the courtship sequence? Can the role of the male *Nasonia* in the reaction chain be mimicked by adept manipulation on your part (e.g., by stroking the antennae of a virgin female with a fine wire)?

Many aspects of *Nasonia* reproductive behavior can be quantified. Enumerate at least five components that can be timed (a stopwatch would be helpful in gathering the data) and several discrete behaviors that can be counted. Your reaction chain can also be made more precise by specifying the relative frequencies of the various possible pathways.

Finally, if pupae of the flies *Musca, Calliphora,* or *Sarcophaga* are available, the final stage of *Nasonia* reproductive behavior may be examined. Following introduction of females into petri dishes containing pupae, you will observe the females' palpation of the pupae, drilling through its surface, and eventually, oviposition. (These behaviors may take longer to elicit than the courtship reaction chain you have just studied—from a few minutes to a few hours.) Again, try to identify the behavioral sequence. Can you design simple experiments that will reveal the means whereby females identify suitable pupae (i.e., what is the role of visual, olfactory, and tactile stimuli)?

REFERENCES

Barrass, R., 1960. The courtship behaviour of *Mormoniella vitripennis. Behaviour* 15:185–209.

Eibl-Eibesfeldt, I., 1970. *Ethology: The biology of behavior.* New York: Holt, Rinehart and Winston.

Lorenz, K., 1952. *King Solomon's ring.* New York: Crowell.

Tinbergen, N., 1951. *The study of instinct.* London: Oxford University Press.

Whiting, A. R., 1967. The biology of the parasitic wasp *Mormoniella vitripennis. Quart. Rev. Biol.* 42:333–406.

PETER MARLER
The Rockefeller University

Courtship Behavior of *Drosophila*

In this exercise you will identify and study the sequence and function of the courtship activities of fruit flies.

Drosophila cultures provide an ideal source of experimental animals that can be used in the analysis of many interesting behavioral problems. Because of their availability and the enormous background of genetic information about them, these flies are being used as subjects of study by many students of behavior.

COMPONENTS OF COURTSHIP BEHAVIOR

The following classification of courtship activities in *Drosophila* is by Spieth (1952). Use this list of descriptions as a basis for initial interpretation and classification but be alert for variations. Watch especially for any aspects of courtship which you think may be subjected to quantitative analysis.

Wing Vibration. In this behavior the male extends one wing (or, in some species, perhaps both wings) laterally from the resting position and then moves the wing or wings rapdily up and down. The lateral displacement varies from 3° to 90°, depending on the species; the vane of the wing is held parallel or almost parallel to the substratum and then vibrated rapidly up and down. The extent of the vertical displacement varies from very short movements to those of considerable amplitude. The speed of vibration is variable and within limits seems to be species-specific. Thus, vibrating consists of two distinct components: lateral displacement, and short, rapid up-and-down movements that occur while the wing is held at the point of maximum displacement. Vibrating occurs in pulses of movement, each one lasting between a fraction of a second and a few seconds. Between each burst of movement the wing is typically returned to the resting position.

Flicking. This is a wing movement of the courting males of some species. The wing is moved quickly out-and-back from the resting position. The vane is typically held parallel or almost parallel to the substratum. Often the complete movement is repeated several times in quick succession. Flicking and vibrating are sometimes difficult to separate, since vibration is essentially flicking with the added vertical movement of the wing. In those species in which this vertical movement is extremely small, it is difficult to determine under low magnification whether the wing is being "vibrated" or "flicked." In flicking, as in vibrating, a few species turn the vane of the wing perpendicular to the substratum, with the posterior edge of the wing directed downward.

Waving. One wing of the male is slowly spread outward from the body to 90°, held in this position, and then relaxed without vibration.

Scissors movement. A courting male, during the interval between wing vibrations, will sometimes open and close both wings in a scissors-like movement. This is rarely seen and appears most typically in some specimens of *D. melanogaster* and its relatives. Apparently, highly excited males display this motion most often, and it may merely be part of the readjustment of the wings to the normal resting state rather than a courtship action.

Fluttering. Nonreceptive females of some species, and usually males of the same species when courted by a male, engage in an inconspicuous but definite and distinct movement of both wings which serves as an effective repelling movement. The wings are slightly elevated, separated from contact with each other, and then moved laterally a tiny amount and vibrated rapidly. In some species, the vibration is vertical, but in others it is horizontal. In most species all movements are small, but in *D. buskii* the wings are spread about 30° and the vertical movements are of considerable amplitude.

Tapping. The male initiates courtship with a foreleg motion, partially extending and simultaneously elevating one foreleg or both forelegs and then striking downward, thus bringing the ventral surface of the tarsus in contact with the individual he is courting. Tapping may occur at various times in the courtship and, as far as Spieth has observed, no courtship is initiated without at least one tapping movement on the part of the courting male.

Licking. The male of many species of *Drosophila,* when he is courting another individual, positions himself closely behind the courted individual, extends his proboscis, and licks the genitalia of the other insect. The duration of contact may be short or prolonged but always involves the labellar surfaces of the courting individual and the genitalia of the courted specimen.

Circling. After posturing at the side or rear of a nonreceptive female, a male will leave his posturing position and quickly circle about the female, facing toward her as he moves. Sometimes he moves until he is in front of her and then retraces his path to the rear; at other times he moves completely about her in an arc of 360°. Often the male engages in special wing or proboscis posturing movements as he circles. Males of some species pause in front of the female during the circling and engage in special posturing actions. Certain species, it should be noted, posture exclusively in front of the females. Circling probably originated as a maneuver to prevent nonreceptive females from escaping from the male's attentions.

Stamping. Sexually excited males may stamp their forefeet, although this does not occur consistently.

Extruding. The nonreceptive female of many species appresses the vaginal plates, contracts certain of the abdominal muscles, and apparently relaxes muscles that are attached to the vaginal plates, and perhaps relaxes the muscles of the vagina as well. This causes the vaginal plates to be carried posteriorly, the displacement possibly being due to a stretching of the articulating membrane that lies between the vaginal plates and the preceding sclerites. Such a complex action results in the formation of a temporary tubelike structure. Every possible degree of extrusion exists, from none in some species to slender, very elongated tubes in others. The extrusion action may be accompanied by other movements, such as directing the tip of the abdomen toward the male's head or elevating the tip of the abdomen high into the air.

Decamping. Nonreceptive females often attempt to escape the male's overtures by running, jumping, or flying away from the immediate vicinity of the courting male. Such movements are categorized under the convenient term, "decamping."

Depressing. Nonreceptive females may prevent mating by curling the tip of the abdomen downward toward the substratum and keeping the wings together. Some species do not curl the abdomen but merely depress the tip. Depressing involves the wings as well as the abdomen. When species that have long wings depress the abdomen, they also depress the wings and hold them firmly in place. Courting males are often unable to get underneath the rigidly held wings when attempting to lick the female's vaginal plates.

Ignoring. The nonreceptive female sometimes, when courted, simply keeps on with whatever activity she has been engaged in, her behavior unchanged by the male's actions. Sometimes, if feeding or preening, she ceases these actions and simply sits quietly.

Countersignaling. In most species of *Drosophila,* males court other males as well as females. A courted male may engage in a series of movements, such as spinning about to face the suitor and then striking him with the forelegs, or he may combine fluttering with kicking. Such movements seem to induce cessation of courtship. All of these sundry movements can be considered as countersignaling. Not all males countersignal, and the apparent effectiveness of the action varies considerably between individuals and especially between species, some of which do not countersignal.

METHODS

Subjects and Materials

Obtain several species of *Drosophila,* keeping sexes and age-groups separate. Males of *Drosophila* may be recognized by the somewhat smaller size, the black-tipped abdomen, and particularly the sex comb on each of the front pair of legs, which looks like a black spot halfway down the leg (Fig. 20.1). Sexing can be readily done shortly after eclosion, while the flies are under light anesthetization with ether.

Apparatus needed for this exercise includes a fly-transferring pipette (Fig. 20.2,*a*), a holding jar (Fig. 20.2,*b*) and an observation chamber (see Fig. 21.1). Learn how to use this equipment before starting your observations.

Procedure

The procedure to be described should be followed faithfully, without any unauthorized modification, because unless everyone uses the same method, it will be difficult to trace the causes of variation in the results.

A. Identification of Components of Courtship Behavior Introduce two females and two males of *D. melanogaster.* It is essential to use care in drawing the flies up into the pipette and in gently blowing them into the observation chamber. If a fly is damaged in the transfer, grooming rather than reproductive behavior will result.

First, observe with the naked eye. Watch for the initial response of the male upon encountering a female. Then bring the dish under a low-power dissecting microscope. Try to identify the behavioral actions listed. Be extremely patient! If nothing happens, try other pairs. Continue until you become able to identify some of the listed activities.

Observe the response of the female to the male. Identify and describe any rejecting behavior by the female toward the male.

B. The Sequence of Courtship Behavior Before proceeding to use other materials, remember that your observation must begin immediately upon introducing the flies into the dish. Do not delay a second in starting to note their responses upon the first encounter.

Introduce two virgin females and two males of any age of *D. melanogaster.* Restrict your observations to the male. Try to record the sequence of occurrence of those behaviors in the male associated with courtship. Use shorthand: WV—C—S, and so on. Make repeated recordings of the sequences of courtship behavior of these flies so that later these data can be treated statistically.

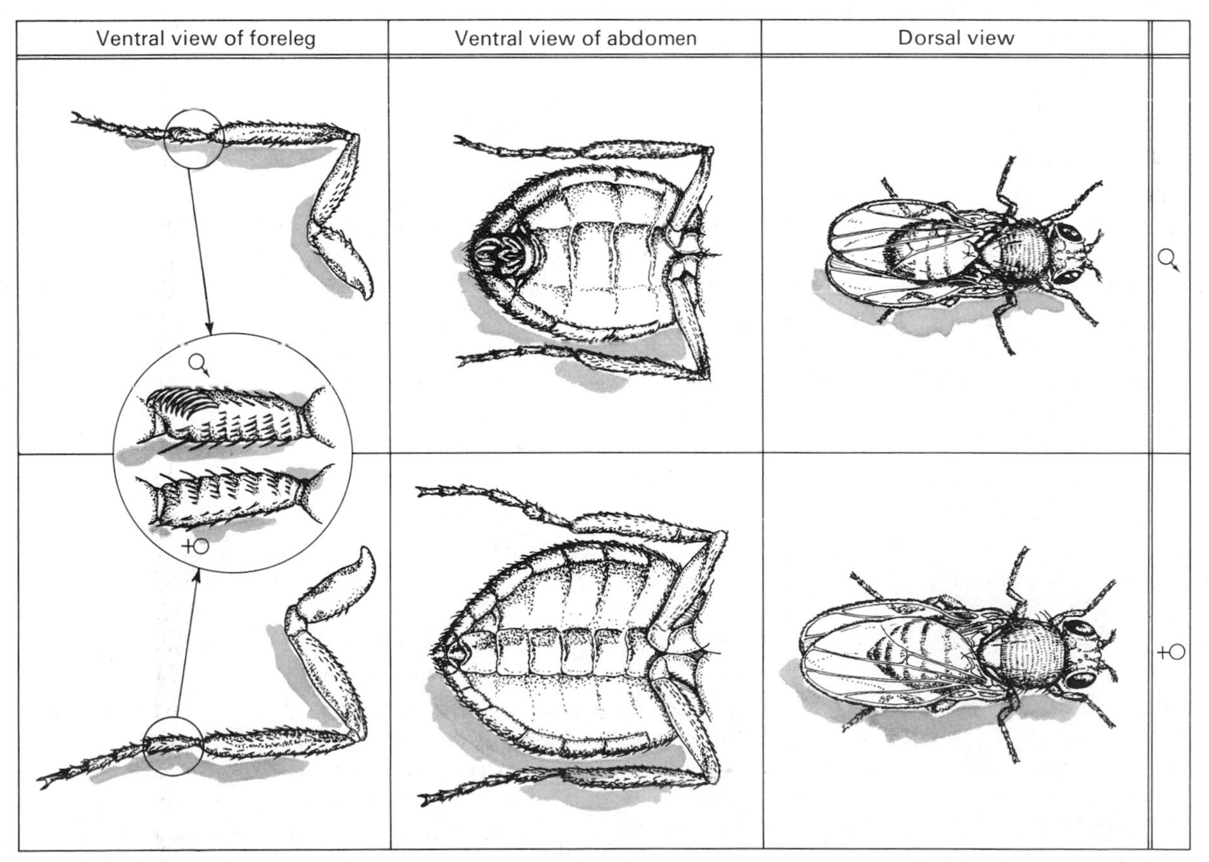

Figure 20-1. Comparison of male and female fruit flies. The further enlargement of forelegs shows the sex comb, a characteristic of male flies. (After Biological Sciences Curriculum Study, 1963.)

Following the same procedure as above, observe the courtship behavior of another species. Notice any behavioral differences between this species and *D. melanogaster*. Describe the nature of the difference. Establish the sequential relationship of the actions as you have done with *D. melanogaster.*

If time and supplies permit, try additional species.

DISCUSSION

Your data should include some long sequences of actions observed in both male and female. An analysis of these may permit you to determine what effects the behavior of one sex has on the other. How far can you go in applying notions of appetitive and consummatory behavior to such analysis?

QUESTIONS

1. On the basis of what you have seen, is there any evidence indicating an exchange of signals between males and females?
2. Which sex takes thc most active role?
3. Do males compete for females?
4. How would you proceed to determine the sensory modalities used in signaling?

ADDITIONAL STUDIES

Observe the effect of amputating the antennae of the male close to the head. Perform the operation a day before the experiment. Introduce two altered males and two virgin females of *D. melanogaster* and watch the behavior of the males. Record any departures from the normal patterns of behavior previously seen. The same experiment can be done with headless flies.

Kill a virgin female and a male *D. melanogaster* by pressing on the thorax and place them in the observation dish 1.5 cm from each other. Introduce a *D. melanogaster* male and see his response to the dead flies. Count separately the number of approaches, courtship actions, and copulation attempts with the dead flies.

Introduce two *D. melanogaster* males and three virgin *D. virilis* females. Note any differences in the male courtship from that seen in the conspecific cross that you observed before. Record the sequential relationship among the classified behavior patterns. Describe how the females reject the males. Do they use the same method here as that used against their conspecific males? Try a reciprocal cross, with *D. melanogaster* females and *D. virilis* males, and perform the same experiment.

Replace the *D. virilis* females with three virgin *D. similans* females, and record as above. Make a reciprocal cross (two *D. similans* males and three virgin *D. melanogaster* females) and take similar data.

In Exercise 21 you will have the opportunity to observe whether the differences in courtship patterns between species reduce the probability of hybridization between species.

REFERENCES

Biological Sciences Curriculum Study, 1963. *High school biology: Green version, student's manual.* Chicago: Rand McNally.

Ehrman, L., 1964. Courtship and mating behavior as a reproductive isolating mechanism in *Drosophila. Amer. Zool.* 4:147–153.

Spieth, H. T., 1952. Mating behavior within the genus *Drosophila* (Diptera). *Bull. Amer. Mus. Nat. Hist.* 99:395–474.

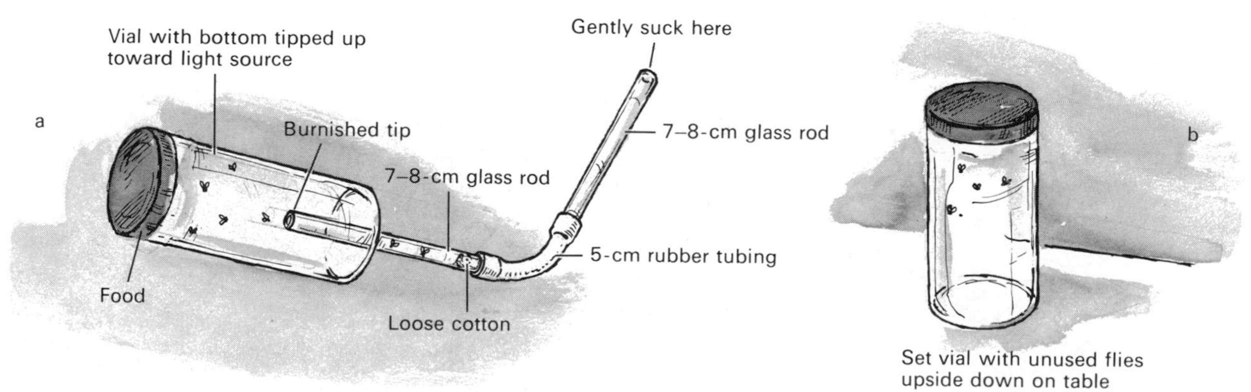

Figure 20-2. Method of transferring flies: (*a*) transfer pipette, (*b*) method of temporarily holding excess flies.

LEE EHRMAN
State University of New York, Purchase

21

Reproductive Isolation in *Drosophila*

Reproductive isolating mechanisms are the instruments by which genetic materials are shaped into discrete arrays known as species. The perfection of reproductive isolation is speciation: i.e., splitting up of the gene pool of a single species into two or more derived pools. Reproductive isolation is also responsible for maintaining the separation of the gene pools of distinct species.

Sexual (or ethological or behavioral) isolation is a most efficient mechanism. It prevents wastage of gametes and of food and space for developing inferior or sterile hybrids. When this process is operative, potential mates of different species do not mate when they meet because the mutual attraction is less than that between conspecific males and females.

Within a species, there may exist preferences for some mates rather than for others. In the *Drosophila* species to be used in this exercise, the choice of mates is largely if not entirely the province of the female sex. *Drosophila* females are able to discriminate between different mutants, between the bearers of different chromosomal inversions, between members of different geographic races, and between members of different species.

In this exercise you will observe mating behavior when two species are placed together. These will be species whose behavior you may have observed individually in the preceding exercise. You will first determine the extent to which reproductive isolating mechanisms are operating. You then should compare the degree of successful isolation with the divergence in mating behavior that you observed previously in these species.

METHODS

Subjects and Materials

Table 21.1 describes several species that may be used in this exercise. Obviously, only a few of the hundreds of *Drosophila* species available for study are mentioned.

Elens and Wattiaux (1964) have described a simple apparatus for measuring the sexual preferences of different mutants of *Drosophila melanogaster* by direct observation. This is essentially an inexpensive glass and wood sandwich (Fig. 21.1) but any transparent box or container with a suitable opening for introducing flies and a grid floor will suffice.

Procedure

Flies introduced into the observation chamber must have been aged to sexual maturity in isolation from individuals of the opposite sex to assure virginity. If the two experimental strains are not otherwise distinguishable, mark one by notching the distal margin of one wing with needles so that the individuals of this strain can be identified with a four-power hand lens or a simple dissecting microscope. It is best to mark the insects while they are anesthetized for separation of the sexes shortly after eclosion. (They will perform better under experimental conditions if not anesthetized repeatedly.) The strain marked with wing notches should be rotated so that no one strain is always marked. (No experimental *Drosophila* should be etherized within 24 hours of their utilization for observation.) Wing notching is, of course, unnecessary for easily distinguishable mutants. For techniques of sexing and transferring flies from stock bottle to observation chamber, see Exercise 20.

Introduce a total of 40 flies, 10 pairs of each of two kinds, into your observation chamber. It is suggested that the flies be introduced from four separate vials, females first, in the following order:

1. 10A Females 2. 10B Females
3. 10A Males 4. 10B Males

It is unlikely that all of the females will mate while under observation. If 50 percent or more of them do, the experimental run may be considered a success.

Equal numbers of each of the four kinds is not always the best choice, and it is interesting to study how the frequency can affect mating success. So although the "10 of each × 4 = 40" ratio should be tried first, other proportions such as the following may be used:

18A Females:2B Females
18A Males:2B Males

or

15A Females:5B Females
15A Males: 5B Males

Make the reciprocals also; a species that was rare in one test should be common in another.

Table 21-1. Subjects.

Species	*Possible types competing for mates*	*Hints*
1. *Drosophila melanogaster**	Mutant (e.g., white-eyed) versus nonmutant (+ / +, wild type) or mutant (e.g., forked) versus mutant (bar-eyed).	This species is the easiest to handle in the laboratory and the number of mutants available is greatest.
2. *D. melanogaster* versus *D. simulans**	Different strains (collected at different sites) of one species versus different strains of the other.	Compare the frequencies of intraspecific homogamic (*D. melanogaster* × *D. melanogaster* or *D. simulans* × *D. simulans*) matings with the interspecific heterogamic ones (*D. melanogaster* × *D. simulans*). Consider the species of the female and that of the male partner in the case of interspecific matings, e.g., *D. melanogaster* female × *D. simulans* male or the reciprocal, *D. simulans* female × *D. melanogaster* male.
3. *D. pseudoobscura***	Strains collected at different localities. There are also some good mutants available, e.g., orange-eyed.	Because the mounts are very brief in this species—i.e., 3 to 4 minutes—they will occur rapidly; for this reason, as soon as both sexes have been transferred into the chambers, pay constant attention or some matings will not be scored.
4. *D. pseudoobscura* versus *D. persimilis**	Same as 2.	
5. Members of the *D. willistoni*† species group	*D. willistoni, D. tropicalis, D. equinoxialis, D. insularis*, and *D. paulistorium*. Any set of these as a demonstration of strong sexual isolation without accompanying morphological differences.	Wing clipping for ease of identification is essential here; rotate the marks used for identification between the strains to be identified. Mutants are not routinely available.

*Collect and separate the sexes at least every 8 hours to ensure virginity, age for 2-3 days.
**Collect and separate the sexes at least every 24 hours to ensure virginity; age for 4 days.
†Collect and separate the sexes at least every 4 hours to ensure virginity; this brief span will make the *D. willistoni* flies more difficult to collect in sufficient numbers than those of the other species listed. However, they are the easiest to record while they are in the observation chambers because a mount will last an average of 17 minutes. Age for 1 week.

As you observe matings in the chamber record these five facts:

1. The time (from the beginning of observations) each mating takes place.
2. Its sequence among other copulas which occur.
3. Where in the chamber the mating pair is located (for this purpose a grid is made on the floor so that each individual mating is counted only once).
4. The kind of female involved.
5. The kind of male involved.

It may also be desirable to record the duration of each mount.

After sufficient data have been accumulated, a "coefficient of joint isolation" may be calculated. (This will apply only to those tests in which equal numbers of each type are introduced.) Briefly, if p_{11}, p_{22}, q_{12}, and q_{21} are the proportion of matings observed between males and females of type 1, males and females of type 2, males of type 1 and females of type 2, and males of type 2 and females of type 1 respectively, then the coefficient equals:

$$p_{11} + p_{22} - q_{12} - q_{21}.$$

For example, if a total of 80 matings were observed and 20 were between males of type 1 and females of type 2, q_{12} would be $20/80 = 0.25$.

Random mating (no preferences, no sexual isolation) will result in a coefficient of zero, and complete isolation (no mating between unlikes) produces a coefficient of $+1.00$.

(The variance $= 4pq/N$ where $p =$ the proportion of homogamic matings $p_{11} + p_{22}$, $q =$ the proportion of heterogamic matings $q_{12} + q_{21}$, and $N =$ the total number of matings observed.)

It is suggested that N be at least one hundred. (For *D. melanogaster,* this would mean running 5 replicates and would take some two hours.)

DISCUSSION AND QUESTIONS

Now that you have identified those crosses that produce the greatest degree of reproductive isolation, compare your findings with the behavioral observations you made in Exercise 20. Do you find evidence that the isolation could arise from divergence in mating patterns? If time permits, for confirmation observe again the behavior patterns in the separate species. Do you notice differences in "athletic ability," i.e., do certain types of males mate more often than others, and do certain types of females accept males more easily than others? Do thresholds differ in the courtship–mounting–insemination sequence?

ADDITIONAL STUDIES

Consider the influence of the relative numbers (common *vs* rare) of competing males on their success in acquiring females; relate to the performance of the competitors when all are equally represented. Is there frequency-dependent mating success within a

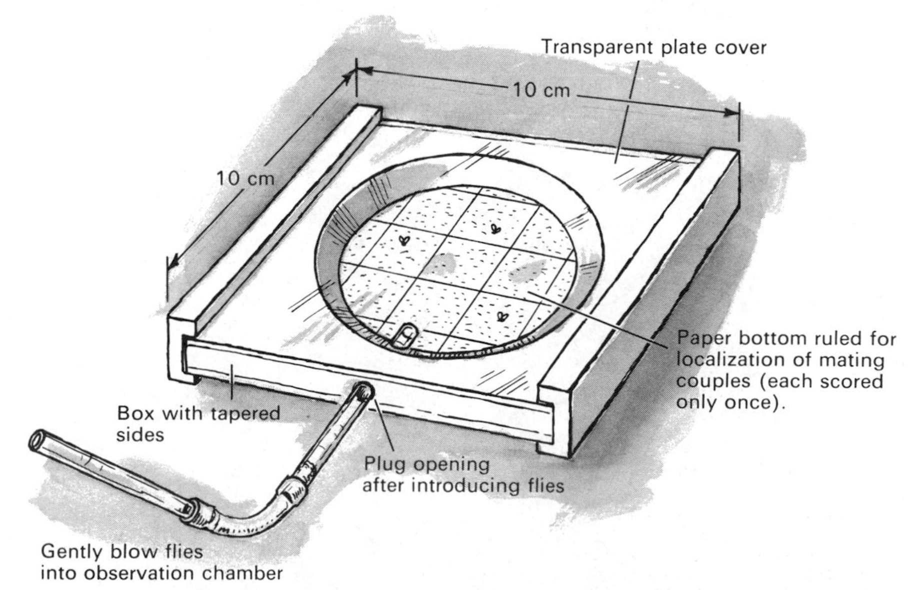

Figure 21-1. Observation chamber for scoring matings.

given species? Between two different species? Does the degree of commonness or rarity (9:1 or 4:6) affect this outcome? If mutants routinely do more poorly than nonmutants, can this outcome be altered by changing the numerical relationship between mutants and nonmutants? What if the environment in the chamber is altered: e.g., by introducing a strange odor? (The behavior in a clean, fresh chamber must be recorded first.) Do any females mate more than once? Any males? Under what conditions? What is the influence of time? When do most matings take place? Do more homogamic matings occur first, or among the late maters? What if parts of the body of individual flies (e.g., legs, antennae, wings) are prevented from functioning normally?

REFERENCES

Dobzhansky, T., 1970. *Genetics of the evolutionary process,* pp. 311–350. New York: Columbia University Press.

Ehrman, L., 1962. Hybrid sterility as an isolating mechanism in the genus *Drosophila. Quart. Rev. Biol.* 37:279–302.

Ehrman, L., 1965. Direct observation of sexual isolation between allopatric and between sympatric strains of the different *Drosophila paulistorum* races. *Evolution* 19:459–464.

Elens, A. A., and J. M. Wattiaux, 1964. Direct observation of sexual isolation. *Drosophila Information Service* 39:118–119.

WILLIAM H. CALHOUN
University of Tennessee

22

Courtship and Mating in Japanese Quail

Courtship patterns unique to a species serve several basic functions, including synchronizing the behavior of the sexes, reproductive isolation of the species, and reduction of intraspecific aggression. Courtship and mating patterns in the Japanese quail, *Coturnix coturnix japonica,* provide a suitable example (Fig. 22.1). These birds are hardy, grow rapidly, and have a strong sexual motivation. If males and females are housed apart prior to the normal breeding season, courtship and mating occurs within minutes after pairing a male and female.

Courtship may be defined as the behaviors occurring prior to the male's mounting of the female, and mating as the behaviors from mounting through ejaculation by the male. Agonistic behavior (threat and attack) is frequently seen as part of the courtship behavior in this species; the male approaches the female (or another male) in a "threat" posture, and the response of the opponent determines whether courtship or fighting takes place. Students should be familiar with the agonistic behavior of the species before undertaking this exercise. When a male is paired with a female, the response of the female usually determines the outcome of the encounter. An aggressive counterattack by the female rebuffs the male, and ends the courtship sequence. If the female responds submissively by assuming a crouch, courtship proceeds. In male-male encounters submissive posturing by the subordinate male sometimes releases mounting by the dominant male.

There are three objectives in this exercise. The first is to identify and describe the components of agonistic behavior exhibited in male-male encounters (see Exercise 2 for the steps involved). The second objective is the identification and description of behaviors occurring during courtship and mating (male-female encounters). The third is quantification of these behaviors and determination of the sequence in which they normally occur.

METHODS

Subjects and Materials

Test birds should be purchased at least one month in advance of the scheduled laboratory period. House all birds singly for 20–30 days before testing. Each group of students (3–6 individuals) will need a stopwatch and materials for data recording. A circular or semicircular enclosure, about 1 meter in diameter, constructed of wire mesh or clear plastic will permit observations from the sides as well as from above. A wall height of 0.5–0.75 meter contains the birds if *one* wing of each bird is clipped. The floor should be covered with sawdust to facilitate cleaning. For easy identification, mark each bird with a colored leg band or by painting a spot on the head or back with a colored felt pen.

Procedure

A. Identification of Behavior Components Place two strange males (ones that have been housed separately from each other) in the enclosure and identify the components of agonistic behavior in both the dominant and subordinate bird. (You will see similar behavior by the male and female during courtship.) After 5 to 10 minutes of observation of one pair, remove the subordinate male and introduce another strange male. Continue the sequence until you are familiar with the agonistic behavior of this species. Two to four pairings are usually adequate.

Next, place a female in the empty enclosure and allow her 5 to 10 minutes to adapt to the strange surroundings. Then, introduce a male in the enclosure with her. Note the female's distinguishing color and posture and contrast it with the male's. At first, just watch the pair, but as you become familiar with their behavior, begin recording the behavior patterns observed using your own descriptive names as needed: avoid names that imply causation or function (see Exercise 1).

For purposes of this exercise, a mating will end with ejaculation (which may be hard to see at first). A copulatory period, composed of several matings, will end when a 5-minute intercopulatory interval (time of last ejaculation to initiation of a new mating) is exceeded. Remove the first pair of birds from the enclosure once the copulatory period, as defined by the 5-minute interval, has ended. Depending on the number of birds and the amount of time available, repeat the observations with untested pairs.

After completing the observations, reassemble for class discussion. Ask representatives from each group to read their lists of behaviors and record them on a blackboard. Through discussion, agree upon a complete list of behaviors and brief descriptive names for each. Make sure this list describes the mating pattern for both sexes. Your instructor may suggest adding behaviors you might have overlooked or which did not occur in the brief period of observation. Next, from this list decide on the most readily recognizable behavior components observed and establish criteria for their identification. It is important to adhere

Figure 22-1. Some behavioral postures of male Japanese quail. (After Farris, 1967.)

rigidly to these criteria in all subsequent observations, or the quantitative phase of the study will fail. Prepare a master check list for subsequent observations.

B. Quantifying Behavior Components During a second observation period, observe a naive pair (male and female) of birds until the 5-minute intercopulatory interval has been exceeded. Record the frequency and duration of occurrence of the behavior components on the master check list. Note recurring sequences of behaviors that might be meaningfully grouped into a larger category, e.g., courtship. If possible, study additional pairs as time and number of birds allow. Lastly, prepare a profile of behavior for each sex, indicating relative frequency or duration of the behaviors. Comments about the sequence of behavior will aid in the description of the courtship and mating behavior.

DISCUSSION

Courtship patterns vary greatly with species; some patterns are prolonged and ample opportunity for careful observation is available. Because *Coturnix* have a relatively short cycle of courtship and mating, repeated observations are needed to establish the complete pattern. Consult Exercises 19, 20, and 28 for further study of courtship and mating behavior.

QUESTIONS

1. What distinct behavior components constituted courtship?
2. Did some behaviors always or usually follow others? What were they?
3. What happened if the female rebuffed the initial approach of the male?
4. What did the crouched, fluffed posture by the female indicate? Did either male in the male-male pairing assume this posture? What did it indicate in that situation?
5. The male was probably rather noisy. Did the female make any sounds? When? What role might this have?
6. Briefly describe the typical sequence of events from initial action by the male through ejaculation. Show how the birds' behaviors interlocked, as in the stickleback studies by Tinbergen (1952). Discuss the observed behavior in terms of Tinbergen's four functions of courtship and mating behavior.

REFERENCES

Aronson, L. R., 1949. An analysis of reproductive behavior in the African mouthbreeding fish, *Tilapia macrocephala. Zoologica* 34:133–155.

Farris, H. E., 1967. Classical conditioning of courting behavior in the Japanese quail. *Coturnix coturnix japonica. J. Exp. Anal. Behav.* 10:213–217.

Stokes, A. W., 1963. Agonistic and sexual behaviour in the chukar partridge. *Anim. Behav.* 11:121–134.

Tinbergen, N., 1952. The curious behavior of the stickleback. *Sci. Amer.* 187(6):22–26. (Offprint 414.)

Wiepkema, P. R., 1961. An ethological analysis of the reproductive behavior of the bitterling (*Rhodeus amarus* Blach). *Behaviour* 16:103–199. [Reprinted in part in S. C. Ratner and M. R. Denny (Eds.), 1964. *Comparative psychology.* Homewood, Ill.: Dorsey Press.]

JAMES H. REYNIERSE
Hope College

Parental Behavior in Cichlid Fish

Cichlid fishes are chiefly native to Africa, Central America, and tropical South America. In their natural habitats cichlids are often territorial, particularly during the breeding season, directing aggressive attacks to individuals of their own as well as other species.

Several species of cichlids breed easily in aquaria. The onset of breeding is signalled by the prominence of extended genital papilla (breeding tubes) and bright courtship colors. A breeding site, usually a flat stone or shallow depression in the gravel, is selected and cleaned, and ritualized courting occurs. As spawning approaches all species become aggressive, driving off intruding fish and defending the nesting site. The parental-care cycle consists of three distinct stages or periods: (1) care of the spawn or eggs; (2) care of the wrigglers, or young fry; and (3) care of the free-swimming older fry. Brood care involves transfer of the fry to brood pits dug in the gravel, retrieval of strays, keeping the fry in a compact school and offspring defense. Mouthbreeders are particularly interesting since they carry the eggs and young in the mouth initially, and later when danger threatens.

The objective of this exercise is to observe and quantify various aspects of parental behavior in cichlid fish during each of the three stages of parental care.

METHOD

Subjects and Materials

Jack Dempseys (*Cichlasoma biocellatum*), Rainbow or Red-eyed cichlids (*Herotilapia multispinosa*) and Mozambique mouthbreeders (*Tilapia mossambica*) have been used in this exercise with success. Both the Jack Dempsey and Rainbow cichlids are aggressive; the Mozambiques tend to be more docile.

The basic apparatus consists of one or more 20–30 gallon aquaria equipped with air pumps, filters, and heaters. The bottom of all aquaria should be covered generously with coarse natural gravel to a depth of about 5 cm. Artificial plants and rock slabs should be present in all aquaria.

Procedure

Two pairs of cichlids should be maintained in each aquarium. Become familiar with individual fish, identifying specific behaviors associated with agonistic encounters, courtship, and parental care. Describe observed behavioral patterns in objective terms, avoiding functional descriptions and names (see Exercise 1). Then concentrate on observing pairs during different stages of the reproductive cycle and describe the parental-care characteristic of each stage. What changes in agonistic behavior (both within and between pairs) occur during the various phases of the reproductive cycle?

Once you have become familiar with the behaviors associated with parental care in your fish, attempt to quantify the more discrete behaviors observed. If you have time, observe one or more pairs every 2 or 3 days through an entire reproductive cycle. Otherwise, observe several pairs (at different stages of their reproductive cycle) during one or two scheduled laboratory periods.

Some behaviors you may want to quantify are:

1. Frequency of head-stand nibbling in which parents pick at decayed eggs or other debris.
2. Frequency and duration of bouts of fanning eggs with pectoral fins (substrate spawners only).
3. Frequency of picking up wrigglers and returning them to the pit.
4. Frequency of retrieving stray free-swimming fry.
5. Frequency of agonistic behaviors (e.g., bites, chases, etc.) directed at mate or neighboring pair.

You will also want to note how the dominance relations and size of defended territory (if any) changes with stage of the reproductive cycle.

If you total your data at frequent intervals (e.g., 5 or 10 minutes) it will be easier to compare scores obtained during different observation periods (which may vary in length).

You may find it helpful to graph your data and conduct statistical analyses when possible.

QUESTIONS

1. Do both sexes participate equally in parental care at all three stages? Identify any differences. Under what conditions did you observe changes in the behavior of the parents toward each other?

2. What is the adaptive significance of the activities observed in each of the three stages of parental care?

Figure 23-1. Aquarium set-up for studying habituation of the agonistic response of a cichlid fish to its mirror image.

ADDITIONAL STUDIES

Using a single pair of cichlids with eggs or fry, you can perform a simple exercise that will dramatically illustrate changes in parental motivation and the appropriateness of the form of parental care as the fry mature. Mount two mirrors against two sides of the aquarium (see Fig. 23.1). For a one-hour session, alternately present the mirrors for two minutes and take them away for two minutes. During each mirror presentation record frequency of "bites"—i.e., open-mouthed contact with the glass side of the aquarium —for both the male and female parent. Other behavioral measures such as fin erection and gill-cover erection can also be recorded. Although it is probably unnecessary to perform this exercise daily, it should be repeated several times during each of the three stages of parental care. Plot the frequency of biting as a function of time during each parental stage. In what way are the behaviors directed to the mirror similar to or different from the activities directed to adjacent pairs of fish in the same aquarium?

For additional information on the social behavior of cichlids consult the following list of references.

REFERENCES

Boer, N. J. de, and B. A. Heuts, 1972. Prior exposure to visual cues affecting dominance in the jewel fish, *Hemichromis bimaculatus* Gill 1862 (Pisces, Cichlidae). *Behaviour* 44:299–321.

Greenberg, B., 1963. Parental behavior and imprinting in cichlid fishes. *Behaviour* 21:127–144.

Heuts, B. A., and N. J. de Boer, 1973. Territory choice guided by familiar object cues from earlier territories in the jewel fish, *Hemichromis bimaculatus* Gill 1862 (Pisces, Cichlidae). *Behaviour* 45: 67–82.

Myrberg, A. A., 1966. Parental recognition of young in cichlid fishes. *Anim. Behav.* 14:565–571.

Weber, P. G., 1970. Visual aspects of egg care behaviour in *Cichlasoma nigrofasciatum*. *Anim. Behav.* 18:688–699.

JOHN S. ROSENKOETTER ROBERT BOICE
University of Missouri at Columbia

24

Chemical Communication in Earthworms

How do earthworms communicate? Visual and auditory stimuli could not carry messages for earthworms since these soil animals have no visual or auditory organs. Mechanical stimuli could be used for communication but this would be limited to a time when worms are in proximity. Chemical stimuli, however, seem to be well suited for earthworm communication. They can be deposited at a particular spot, they can remain active for a long period of time, they can be specific since a myriad of chemical combinations are available, and other earthworms can readily detect them with their sensitive chemoreceptors. The chemicals that are expelled from an organism and elicit a response in a conspecific organism are called pheromones; interspecific chemical messengers are called allomones. This exercise is designed to show that earthworms extrude a substance, pheromone or allomone, that can transmit a message to other earthworms.

METHODS

Subjects and Materials

The subjects of this study are any of various species of earthworms. Gates (1972) gives keys for earthworm identification, but this study does not require that the species be identified as long as different species can be distinguished.

Earthworms should be handled carefully and gently. Either fingers or forceps may be used to grasp worms, and they are easily picked up from a flat surface if a pencil, probe, or forceps is slid under them first.

Recommended materials are forceps, paper towels, scissors, waxed paper, two size D batteries connected in series with a wire attached to each contact, a strong solution of table salt and water, and a ring made by cutting the top 2 cm from a round plastic food container.

Procedure

Cut strips of paper towels 2 cm wide, soak them in the salt solution, and arrange them in a square on a piece of waxed paper. Put two earthworms of the same species in the center of the square (open field) and observe their response to each other, to the waxed paper, and to the salt solution.

Place the ring made from a plastic food container on a different piece of waxed paper, and put another earthworm inside. Shock the worm by touching it briefly with the leads from the two size D batteries. (A shock source set at about 0.05 milliampere may be used instead.) The shock will usually cause the worm to extrude coelomic fluid through its dorsal pores, which are located in the grooves between the segments. The coelomic fluid of earthworms is a clear to yellowish fluid, depending upon the species, that fills each segment of their bodies. It functions as a hydrostatic skeleton and also as a buffer that washes off irritants and protects the epidermis from desiccation.

When the waxed paper has a copious amount of coelomic fluid on it, remove the worm and the ring. Cut the fluid-covered waxed paper into pieces 2 cm square and place a square in front of one of the worms in the open field. Compare the response to that given to a conspecific worm or to the salt solution. Determine whether the response to coelomic fluid is positive or negative: if positive, the subject will stop crawling and maintain contact for several seconds; if negative, the subject will jerk up its prostomium (the sensitive anterior tip) and back away from the stimulus. Then, test the response of the other worm in the open field. Replicate this procedure several times using naive subjects.

Repeat the procedure, testing the response to mucus (secreted during normal activity) instead of coelomic fluid. The only change in the procedure is not to shock the worm that is placed in the ring. A binomial test (Siegel, 1956) can be used to determine whether the frequency of response to coelomic fluid is significantly different from that of mucus.

For a more thorough study, use waxed paper in three conditions: clean, covered with mucus, and covered with coelomic fluid. The same test worm should be exposed to different samples: if to two, analyze your results by a binomial test; if to three, analyze by a chi-square one-sample test (Siegel, 1956).

If you have more than one species of earthworm, first test them intraspecifically (as a pheromone) to determine whether their coelomic fluid is an attractant or a repellant. Then compare the response to coelomic fluid interspecifically (as an allomone) to see if the same message is communicated to other species of earthworms.

QUESTIONS

1. Does the coelomic fluid of each species of earthworm studied possess a chemical messenger? What is the message? Is the message the same for different species?
2. How could earthworms use this form of communication in their natural habitat? (Consider both intraspecific and interspecific uses.)
3. Coelomic fluid is elicited by shock of the sort traditionally used by comparative psychologists in T-maze training. How might the presence of this fluid in the shock area (the "incorrect" arm of the maze) confound these apparent demonstrations of learning?

ADDITIONAL STUDIES

An alternate method for testing a worm's response to chemical substances is to put the worm, anterior first, into a funnel that has a piece of clear vinyl tubing attached to its neck. The funnel orients the worm into the tubing. The end of the tubing can be placed near a substance to be tested. The response, forward through the substance or backward away from the substance, is easily recorded.

A method that gives quick results is simply to shock a worm and then place it in an open field with a test worm. The problem with this method is that it is difficult to state whether the test worm responds to the mucus or the coelomic fluid of the shocked worm.

Question 3 suggests that the presence of pheromones in a T-maze could confound attempts to demonstrate learning. Design an experiment which would compare performance in the presence and absence of such cues.

Earthworm coelomic fluid apparently also functions as an allomone when it acts as a predator repellant. Predators such as moles, salamanders, toads, grasshopper mice, and quail reject earthworms as food if they are covered with coelomic fluid. You may wish to try this effect on any potential earthworm predator that is readily available.

REFERENCES

Gates, G. E., 1972. Burmese earthworms. *Trans. Amer. Phil. Soc.* (N.S.)62(7):1–326.

Laverack, M. W., 1963. *The physiology of earthworms.* New York: Pergamon.

Ratner, S. C., and R. Boice, 1971. Behavioral characteristics and functions of pheromones of earthworms. *Psychol. Rec.* 21:363–371.

Ressler, R. H., R. B. Cialdini, M. L. Ghoca, and S. M. Kleist, 1968. Alarm pheromone in the earthworm *Lumbricus terrestris. Science* 161:597–599.

Rosenkoetter, J. S., and R. Boice. Earthworm pheromones and T-maze performance. *J. Comp Physiol. Psychol.* (in press).

Siegel, S., 1956. *Nonparametric statistics for the behavioral sciences.* New York: McGraw-Hill.

Whittaker, R. H., and P. P. Feeny, 1971. Allelochemics: Chemical interactions between species. *Science* 171:757–770.

Wilson, E. O., 1968. Chemical systems. *In* T. A. Sebeok (Ed.), *Animal communication,* pp. 75–102. Bloomington: Indiana University Press.

Wilson, E. O., 1970. Chemical communication within animal species. *In* E. Sondheimer and J. B. Simeone (Eds.), *Chemical ecology,* pp. 133–155. New York: Academic Press.

LEE C. DRICKAMER
Williams College

25

Hormonal and Social Influences on the Scent-Marking Behavior of the Mongolian Gerbil

A wide variety of mammals mark their home ranges or territories with scent, either in the feces and urine, or produced by specialized scent-marking glands (Ralls, 1971). Olfactory cues and scent-marking may be used by mammals to communicate social (e.g., sexual, dominance) status, to facilitate individual and group recognition, and to label the habitat for use in orientation (Eisenberg and Kleinman, 1972; Johnson, 1973). Olfactory communication is indirect: the sender and receiver need not be present simultaneously. Instead, an animal deposits an olfactory cue and the message is potentially received by all conspecifics that encounter the scent mark.

In this exercise you will study the marking behavior of the male Mongolian gerbil (*Meriones unguiculatus*). Female gerbils also mark, but with a much lower frequency. Adult gerbils possess a ventral sebaceous gland which is used to mark objects in the environment (Thiessen, 1973). The behavioral act of marking consists of a noticeable lowering of the body over the surface of an object, a momentary pause in forward movement and a variable number of rubbing movements during which sebum, the substance produced by the gland, is deposited on the surface of the object being marked.

The objective of this exercise is to investigate hormonal and social factors influencing the marking behavior of the adult male gerbil. First, you will determine whether there is a relationship between the size of the ventral sebaceous gland and the frequency of marking behavior. Since the circulating level of male hormone, testosterone, influences the size of the ventral sebaceous gland (Thiessen et al., 1968) the relationship between gland size and marking provides an indirect assessment of the effect of male hormone on this behavior. Then you will determine whether previous marking by another gerbil influences the amount and location of marking behavior. Finally, you will determine the extent to which marking is influenced by the simultaneous presence of several adult male gerbils in the test area.

METHODS

Subjects and Materials

The subjects for this experiment are adult male Mongolian gerbils (5/lab group). The equipment needed for each student group includes:

1. Fine scissors for clipping fur away from the gland.
2. A clear plastic millimeter ruler for measuring the gland size.
3. An observation arena. Each arena should be about 1 meter square with sides about ½ meter high. Masonite or any other easily cleaned surface should be used for pen construction. Nine to 25 evenly spaced pegs or blocks, preferably of Plexiglas, should be affixed to the floor. Each peg should be about 3 centimeters square and should project no more than 1½ centimeters above the floor of the arena.

Procedure

A. Gland Size and Frequency of Marking Grasp the gerbil firmly, but carefully, by the back of the neck and hold the tail between your little finger and the base of your palm. Turn the animal over and part the hair in the center of the ventral abdomen, exposing the elliptical patch of gland tissue. By carefully clipping away all of the hair surrounding the gland you can expose the entire patch for measurement. Using the millimeter ruler measure the length and width of the ventral gland. At least two students in each group should independently obtain these measurements. Then compute the average length and width of the gland. For each gerbil multiply length times width to obtain a crude index of gland size. Alternatively you may wish to use the measurement data in the formula for the area of an ellipse since this is the shape the gland approximates.

Test each gerbil individually in the arena for 10 minutes. During this test period count all occurrences of marking behavior and the total number of different pegs which are marked. After each gerbil has been tested you must carefully clean the marking posts and the floor of the arena with an alcohol solution. Why? When you have tested five gerbils you can then make graphs (or histograms) of the relationship between gland size and the two variables, number of marks in 10 minutes and the total number of pegs marked.

B. Effect of Prior Marking The purpose of this part is to determine whether prior marking by another gerbil affects the rate and location of marks made by a test gerbil. With your arena clean allow a gerbil to mark for 5 minutes, noting which pegs are marked. Remove the animal, but do not wipe the arena. Place a second (test) gerbil in the arena for a 10-minute period and record the frequency and location of all marks. After the second gerbil has been tested clean the arena. Repeat this procedure until four or five male gerbils have been tested.

C. Presence of Other Gerbils This part is designed to determine whether marking behavior is influenced by the presence of other male gerbils. Mark a group of three male gerbils with a fur clip or dye to permit individual identification. Place the three males in the arena. During a 10-minute observation period record the frequency and location of marks for each of the three test subjects. After testing, clean the arena and repeat this process with two additional groups of three males each.

QUESTIONS

1. Does there appear to be a direct relationship between the size of the ventral sebaceous gland and either the total marking frequency or the number of pegs marked? Do you think that clipping the fur around the gland in any way affects the marking behavior of the gerbils? What are some of the possible functions of such scent-marking behavior, other than territorial delineation?
2. Using the data from Part A as a control: Do animals mark more, or less, when another gerbil has previously marked the same arena? Does the test gerbil mark the same pegs as the first gerbil? What is the significance of the pattern of marking behavior observed under these conditions?
3. What changes in marking behavior patterns and frequencies were observed when the group of gerbils were simultaneously placed in the arena? Any aggression? Do the individuals of the group mark more or less than the animals tested singly in Part A? Why? Among the three gerbils in each test group did one mark significantly more than the others? Cite possible reasons.

ADDITIONAL STUDIES

To investigate further the effects of hormones on marking behavior you may wish to study castrated male gerbils. Castration procedures may be found in papers by Thiessen et al. (1968; 1970). After castrating 10 or 12 adult male gerbils allow three weeks for surgical recovery and then divide the group into two subgroups of equal numbers. The first subgroup will remain simply castrates. Test this group, one at a time, in the arena using the procedure of Part A above. Note the reduction in gland size in these castrated males. Is there a corresponding reduction in marking behavior?

Each gerbil in the second subgroup should be given a pellet implant of testosterone propionate. The pellets should be implanted in the back of the neck region under the skin and fascia layers. The dosage of the pellet can also be varied; a 10-mg pellet is suggested as a good starting point. Allow about one week after the pellet implant before testing. Does the hormone replacement treatment alter the size of the ventral sebaceous gland? Test these gerbils in the arena using the procedure of Part A. Compare the frequencies of marking behavior among these gerbils, the castrated gerbils, and the intact male gerbils tested in Part A. Are there any differences? Why?

An alternative subject for this entire experiment is the adult male golden hamster (*Mesocricetus auratus*). Male hamsters have paired lateral flank glands (black exterior) which can be exposed by clipping fur on the sides of the abdomen near the midline and at the level of the rear legs. For hamsters the pegs should be approximately 7 centimeters high. All other procedures are exactly the same as those described for gerbils.

REFERENCES

Drickamer, L. C., J. C. Vandenbergh, and D. R. Colby, 1973. Predictors of social dominance in the adult male golden hamster (*Mesocricetus auratus*). *Anim. Behav.* 21:557–563.

Eisenberg, J. F., and D. G. Kleinman, 1972. Olfactory communication in mammals. *Annu. Rev. Ecol. Syst.* 3:1–32.

Johnson, R. P., 1973. Scent marking in mammals. *Anim. Behav.* 21:521–535.

Ralls, K., 1971. Mammalian scent marking. *Science* 171:443–450.

Thiessen, D. D., 1973. Footholds for survival. *Amer. Sci.* 61:346–351.

Thiessen, D. D., M. Friend, and G. Lindzey, 1968. Androgen control of territorial marking in the Mongolian gerbil (*Meriones unguiculatus*). *Science* 160:432–442.

Thiessen, D. D., S. L. Blum, and G. Lindzey, 1970. A scent marking response associated with the ventral sebaceous gland of the Mongolian gerbil (*Meriones unguiculatus*). *Anim. Behav.* 18:26–30.

Vandenbergh, J. G., 1971. The effects of gonadal hormones on the aggressive behavior of adult golden hamsters (*Mesocricetus auratus*). *Anim. Behav.* 19:589–594.

JAMES H. REYNIERSE
Hope College

Aggregation Formation in Planarians

Planarians form large aggregations both naturally and in the laboratory. Such aggregations, although probably not indicative of social organization, represent a simple and primitive form of social behavior in which visual and chemical characteristics of existing aggregations recruit additional planarians (Reynierse, Gleason, and Ottemann, 1969). In contrast, Fraenkel and Gunn (1961) suggest that aggregation formation may represent only a common response to optimal intensities of light. As applied to photonegative planarians such a light-intensity hypothesis suggests that aggregations should form only in optimal environments which are in shadow. On the other hand, the social hypothesis suggests that aggregations should occur in many environments. The purpose of this exercise is to see how various factors affect aggregation in planarians.

METHODS

Subjects and Materials

For this exercise almost any planarian species will be satisfactory. Colonies of one or two species of planaria should be maintained in bowls or pans containing aged tap water at a cool to moderate temperature (e.g., 15–24°C). If colonies are to be maintained for a long time, give the planarians a weekly feeding of beef liver for about one hour and change the water in the colony bowls immediately after this feeding.

Other materials required are small circular bowls 10–15 cm in diameter, a supply of aged tap water, black electrical tape, a few pieces of cardboard, and a pipette for transferring planarians. Most exercises can be done with diffuse, overhead lighting, but if a variable transformer is available it can provide more precise control of light conditions.

Procedure

Unless specifically noted, each part of the exercise can be completed under normal overhead lighting. If enough planarians are available, several parts of the exercise can be in effect concurrently, with observers monitoring the several populations simultaneously. Although aggregations usually begin to form within 1 or 2 hours, complete aggregation usually requires a longer time. For this reason it is desirable to make at least spot observations of the undisturbed populations during the next 24 hours.

A. Heterogeneous and Homogeneous Illumination Use a pipette to transfer equal numbers of planarians (25–40) from colony containers to each of two small bowls containing about 3 cm of aged tap water. Attach a cardboard cover to the top of one of the bowls so that half of the bowl is directly illuminated while the other half is in shadow (Fig. 26.1). Keep the remaining bowl of planarians uncovered and directly illuminated. Every 5 minutes observe and record the number of planarians in each bowl that are at rest as well as those that are in aggregations. Agree on some arbitrary but consistent criteria for including individual planarians in an aggregation. For example, individual planarians should not be counted as part of an aggregation if they are moving or if more than 1 cm from another individual in the aggregation. But as a rule aggregations are well defined and the aggregation criteria are required only to include or exclude animals on the periphery of an aggregation. Construct a graph plotting the number of aggregated planarians in each bowl as a function of time. Do aggregations form under homogeneous light conditions? Do they form under heterogeneous light conditions? If so, where? What are the implications of these results for the light-intensity and social hypotheses about aggregation in planarians?

B. Contrast and Recruitment Place a 3-cm strip of black electrical tape anywhere (bottom or side) in a small uncovered bowl. Add about 3 cm of aged tap water, completely covering the tape. Introduce 25–40 planarians and observe them every 5 minutes during the laboratory period. After 24 hours have elapsed and a large aggregation has formed, carefully introduce 15 additional planarians of the same species in

Figure 26-1. Aggregated planarians in darkened half of bowl.

the bowl. Did the original aggregation form on the black tape? Do the newly introduced planarians join existing aggregations?

C. Species Differentiation Although the first two parts of this exercise are directly relevant to the light-intensity hypothesis they are only indirectly related to the social hypotheses. If two species of planarians were maintained together in the same bowl, the formation of distinctive aggregations according to species would provide strong direct support for the social hypothesis.

Place 40 planarians, 20 of each of two different species, in a small bowl containing about 3 cm of aged tap water. Observe the planarians throughout the laboratory period, noting especially if planarians form separate aggregations according to species. If several species of planarians are available make several different comparisons using different pairs of species.

QUESTIONS

1. What evidence is there that factors other than light affect aggregation formation in planarians? Which sensory modalities are involved?
2. Do planarians tend to form separate aggregations according to species? Is this characteristic of all the planarian species you examined? Are the aggregations temporary or relatively permanent?

ADDITIONAL STUDIES

There are a number of technical difficulties involved in examining aggregation behavior in a darkened environment. For example, introducing even a dim light to count the number of planarians aggregated may make many stationary planarians active and thereby disrupt existing aggregations. Nevertheless, it is interesting to ask whether or not planarians aggregate under homogeneous dark conditions. Introduce 25–40 planarians into a small bowl containing about 3 cm of aged tap water. Either cover the bowl completely, or, if available, use a dark room. Observe the distribution of individual planarians and record the number of aggregated planarians at 15-minute intervals. What are the implications of your findings for the light intensity and social hypotheses of planarian aggregations?

If you are fortunate, you may have planarians in the laboratory when they are actively reproducing sexually. Under these conditions unusually large aggregations are formed in which aggregation density is very high. These large aggregations are accompanied by clearly visible mucus secretions, and within a few days several reproductive capsules appear (Reynierse, Gleason & Auld, 1969). Examine such aggregations and compare with the aggregations formed under nonreproductive conditions.

Several variations of the contrast and recruitment procedures (Part B) have been reported by Reynierse, Gleason, and Ottemann (1969), including a generalization to heterogeneous illumination and using larger black contrasting surfaces (e.g., covering half the bowl with black tape). If time is available it may be profitable to examine some of these related procedures.

REFERENCES

Fraenkel, G. S., and D. L. Gunn, 1961. *The orientation of animals.* New York: Dover Publications.

Reynierse, J. H., 1966. Some effects of light on the formation of aggregations in planaria, *Phagocata gracilis. Anim. Behav.* 14:246–250.

Reynierse, J. H., 1967. Aggregation formation of planaria, *Phagocata gracilis* and *Cura foremani:* Species differentiation. *Anim. Behav.* 15:270–272.

Reynierse, J. H., and M. J. Scavio, 1968. Contrasting background conditions for aggregation in planaria. *Nature* 220:258–260.

Reynierse, J. H., K. K. Gleason, and R. Ottemann, 1969. Mechanisms producing aggregations in planaria. *Anim. Behav.* 17:47–63.

Reynierse, J. H., K. G. Auld, and M. J. Scavio, 1969. Preliminary note on seasonal production of planarian pheromone. *Psychol. Rep.* 24:705–706.

HUGH DINGLE
University of Iowa

27

Agonistic Behavior and the Social Organization of Crickets

Crickets are excellent animals to use in studies of aggressive behavior since they are active, exhibit a wide repertoire of behavior, and are easy to maintain. Your purpose in this exercise is to identify and study the agonistic behavior and social organization of crickets. Cricket social organization is exhibited in two forms: dominance hierarchies and territoriality. Dominance hierarchies, as the name implies, are formed when group members consistently dominate (i.e., displace) other group members (see Exercises 29, 32, 33). Such hierarchies may be linear, in which A defeats B who defeats C, and so on; or more complex, where A may defeat B and B defeat C, but C in turn defeats A. Territories are areas defended by an aggressive animal. For crickets this is commonly the area around a burrow or an appropriated shelter. Dominance in the territorial context frequently depends on where (in whose territory) an encounter takes place.

METHODS

Subjects and Materials

Several species of crickets are suitable for these experiments. The most active species, which have the most readily determined behavior patterns, are the various field crickets of the genus *Acheta.* They include *A. pennsylvanicus,* the common large black cricket of late summer and fall in northeastern North America; *A. veletis,* a spring form; and *A. firmus, A. fultoni, A. rubens, A. vernalis,* and others. The domestic cricket, *Acheta* (= *Gryllus*) *domesticus,* which can be obtained from several supply houses or perhaps from a local fish-bait dealer, is less aggressive than the field forms and not as suitable for this exercise. Although fights are rare among *A. domesticus,* their song patterns change during aggressive encounters; hence, this species can be used when the others are unavailable. Much smaller crickets of the genus *Nemobius* become very abundant in many areas in the fall. Their small size makes them perhaps less desirable as experimental animals, but nevertheless they show behavior that is characteristic of their larger relatives as well as certain types of behavior specific to themselves. They are certainly acceptable if larger forms are unavailable, and they are useful for comparative studies.

Crickets may be kept in the laboratory on such foods as chicken mash, oatmeal, or cracked corn, and should be supplied either with water from a cotton wick or with occasional pieces of apple or pear. Some field forms have proved difficult to culture because of diapause, although *A. firmus* has been cultured successfully. You can maintain domestic or house crickets, however, on the above diet. For all species, use damp sand for oviposition. The best type of container is a large one filled with shredded paper to provide crevices; do not let the food get damp, and take care to prevent overcrowding. With a minimum amount of care, you can maintain a thriving culture indefinitely. You may mark crickets for individual recognition by painting the thorax with various colors of airplane dope or nail polish applied with a toothpick.

Procedure

A. Aggression and the Dominance Hierarchy
Place three to five marked males (see Fig. 27.1 for sex identification), each of which has been isolated for 24 hours or more before the start of the experiment, in a container such as a 5- or 10-gallon aquarium whose bottom is covered with sand. Begin making observations immediately. (*Note:* Some investigators give animals a varying length of time to "adjust" to new surroundings. This is not recommended because it effectively eliminates a significant portion of the relevant behavior of crickets, and of many other animals as well.) Observe what happens when two animals come in contact. Note any changes in pattern, rate, or loudness of the song. Does a male's response depend on the direction from which another male approaches? Alexander (1961) distinguishes five intensities of encounter between males of various species of *Acheta.* Try making an independent classification of the various encounters you observe and then compare your results with Alexander's, paying particular attention to the criteria used to set up the classifications. Are the classifications a function of the behavior or do they represent arbitrary impressions of the observer? Record wins and losses during encounters.

If you have taken careful notes, you can use your data to determine whether or not a dominance hierarchy was present in the group of crickets. From the results of encounters calculate a "won and lost record" for each individual. Can you demonstrate from these data that a dominance hierarchy is present? Can you tabulate your data in such a way that provides more information than just a simple frequency distribution of wins and losses?

If time allows, repeat these experiments, using females instead of males.

1. How long does it take a stable hierarchy to become established? Is it linear or complex?

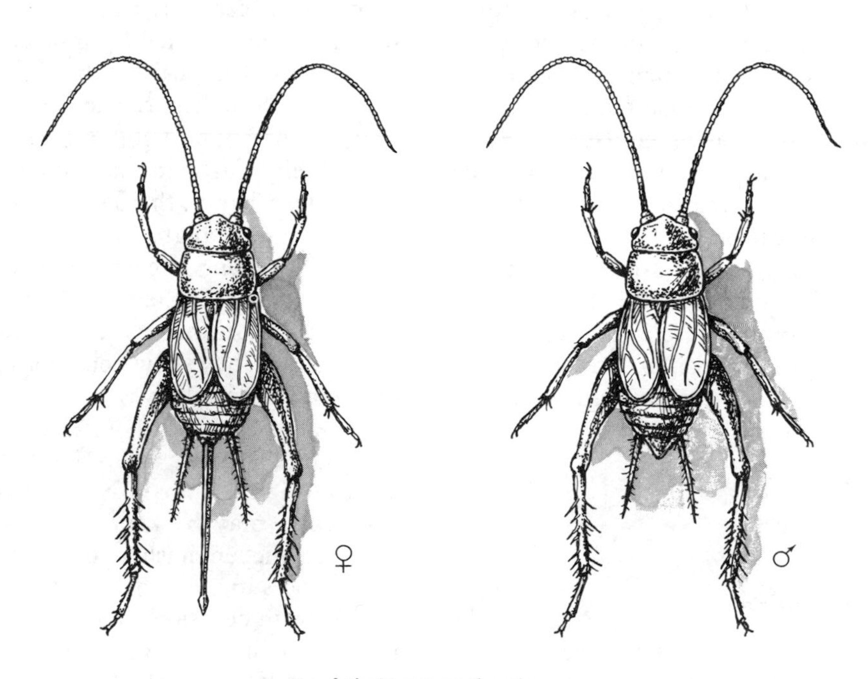

Acheta pennsylvanica
Figure 27-1. Female and male crickets: dorsal view.

2. Does the frequency and intensity of encounters vary with position in the hierarchy?

3. How do factors such as age, size, or physiological state (e.g., recent copulation; see Exercise No. 28) affect dominance?

4. Can you think of any other experiments that might provide insight into the phenomenon of dominance in crickets?

B. Territoriality Set up several containers, each with a small inverted pillbox or matchbox (one end open) in one corner, and introduce a single male into each (Fig. 27.2). Field crickets, which are burrowers, are especially likely to establish a "home" under such a small box; house crickets will probably do the same if no other shelter is available. Allow the resident animal a few days to become established and then introduce a second male and observe the ensuing behavior.

1. Can you observe any evidence of territorial behavior in the resident male? What form does it take?

2. Where in the container are encounters between the two individuals likely to take place?

Conduct several tests with animals when dominant-subordinate relationships are unknown. Can an established animal successfully defend his "home" against an animal previously dominant over him?

Now set up containers with two boxes at opposite ends and introduce three animals whose dominant-subordinate relationships are known.

1. Who is successful in establishing a "home?"

2. How long does it take for territories to become established?

3. Do encounters take place between the two established males? If so, where? (This point may be more readily determined if the extraneous animal is removed.)

Finally, introduce a female and observe what happens.

ADDITIONAL STUDIES

Alexander has reported that when the antennae of crickets are removed, the lashing behavior that usually takes place with these structures is transferred to the mouthparts. Read Alexander's account

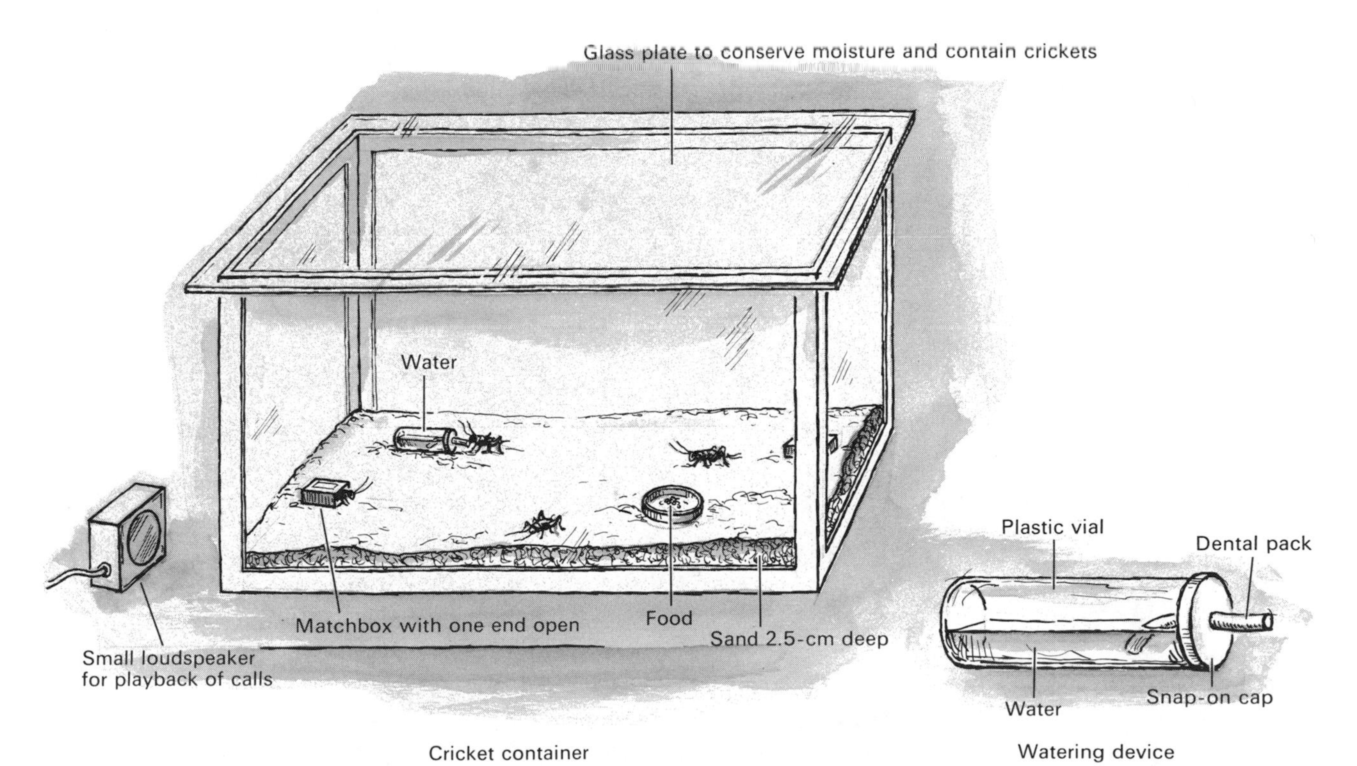

Figure 27-2. Terrarium for observing territorial behavior of crickets.

(in Alexander, 1961), and then observe your animals when the antennae are removed. You may also wish to construct a pair of artificial antennae as illustrated by Alexander and observe what happens when you lash a male cricket with them. If time permits, try combining the artificial lashing with song playback (see Exercise 28). In addition you may wish to try removing other body parts, or blinding the crickets (two coats of partly dried and therefore viscous black airplane dope applied to the eyes are effective, as are various black enamel paints).

REFERENCES

Alexander, R. D., 1957. The taxonomy of the field crickets of the eastern United States (Orthoptera: Gryllidae: Acheta). *Ann. Entomol. Soc. Amer.* 50:584–602.

Alexander, R. D., 1960. Sound communication in Orthoptera and Cicadidae. *In* W. E. Lanyon and W. N. Tavolga (Eds.), *Animal sounds and communication* (Publication No. 7). Washington, D.C.: American Institute of Biological Sciences.

Alexander, R. D., 1961. Aggressiveness, territoriality, and sexual behavior in field crickets (Orthoptera: Gryllidae). *Behaviour* 17:130–223.

Alexander, R. D., and R. S. Bigelow, 1960. Allochronic speciation in field crickets, and a new species, *Acheta veletis. Evolution* 14:334–346.

Alexander, R. D., and E. S. Thomas, 1959. Systematic and behavioral studies on the crickets of the *Nemobius fasciatus* group (Orthoptera: Gryllidae: Nemobiinae). *Ann. Entomol. Soc. Amer.* 52:591–605.

Ghouri, A. S. K., and J. E. McFarlane, 1957. Reproductive isolation in the house cricket (Orthoptera: Gryllidae). *Psyche J. Entomol.* 64:30–36.

HUGH DINGLE
University of Iowa

28

Sexual Behavior of Crickets

The objective of this exercise is to study various parameters of sexual behavior and associated auditory communication in crickets. Exercise 27 discusses several species of crickets suitable for these experiments and general methods for rearing and maintaining them in the laboratory. Read through Exercise 27 before proceeding.

METHODS

Subjects and Materials

The same subjects and apparatus used in Exercise 27 are suitable for this exercise.

Procedure

A. Sexual Behavior Introduce a male and female (see Fig. 27.1 for sex identification) into a container and carefully observe the ensuing sequence of events. (To ensure success it is a good idea to keep the two sexes isolated for a few days before beginning this exercise.) Note especially the initial response of the male to the female.

1. How, for instance, does he know that she is a female and not a rival male?
2. What happens to his song?
3. Does she sing also?
4. What structures are involved in producing and amplifying sounds in crickets? Are there sex differences?

After a period of courtship, the female mounts the male from behind and the male attaches a spermatophore to her genitalia. The latter is a small clear capsule at the end of a short stalk. It can readily be seen, and with a little patience it is not difficult to observe the details of spermatophore transfer. After transfer has occurred there is a short period of post-copulatory behavior. What is the time interval between copulations?

Next, introduce a female into a group of males that have established a dominance hierarchy (see Exercise 27). (It is a good idea to isolate the female for a few days beforehand to increase the chances that she will be receptive.)

1. Which male or males succeed in mating with her?
2. How often does copulation occur, and which males are involved?
3. What effect, if any, does the introduction of the female have on the dominance hierarchy?

B. Nature and Function of Song If a tape recorder is available, it is possible to analyze the cricket song and elucidate more fully its function.

First, record sequences of the different kinds of song you have observed (e.g., calling song, courtship song). Once you have the songs on tape you can play them back to different groups of crickets under different situations. For example, can you design an experiment to determine whether or not the various songs will attract females to the singer? Will they attract males? If such attraction does occur, which songs are the most effective?

You will undoubtedly observe various behaviors occurring while songs are being played back. Can you determine if they are occurring in response to the song? Try putting two females, for example, in a small container and playing back songs of an aggressive and courting male. Do the two songs elicit specific responses from the females? Can you relate the responses to the normal behavior of the animals previously observed? Finally, can you ascertain what it is about the song–frequency, rate, pattern, and so on–that conveys most of the information?

REFERENCES

Haskell, P. T., 1953. The stridulation behaviour of the domestic cricket. *Brit. J. Anim. Behav.* 1:120–121.

Khalifa, A., 1949. The mechanism of insemination and the mode of action of the spermatophore in *Gryllus domesticus. Quart. J. Microsc. Sci.* 90:281–292.

Khalifa, A., 1950. Sexual behaviour in *Gryllus domesticus L. Behaviour* 2:264–274.

Murphey, R. K., and M. D. Zaretsky, 1972. Orientation to calling song by female crickets, *Scapsipedus marginatus* (Gryllidae). *J. Exp. Biol.* 56:335–352.

See Exercise 27 for additional references on cricket behavior.

DAVID C. NEWTON
Central Connecticut State College

Social Organization in Crayfish

Social behavior is shown by animals that aggregate and respond to conspecifics. Animals showing such behavior may be organized in social hierarchies in which dominant individuals may take precedence over subordinates in feeding, territory selection, mating, and other activities. Dominance is frequently established after a series of ritualized encounters that, under natural conditions, seldom harm members of the species. Once a hierarchy is established, brief encounters reinforce dominant or subordinate rank positions. Thus, energy expended in aggressive encounters is minimal within the established hierarchy.

Many organisms maintain a territory from which other members of their own species are excluded. The size of the territory varies, depending on the kind of animal and the physical attributes of the animal's preferred habitat. Aggressive encounters between neighbors at territorial boundaries disperse the animals in the available habitat and frequently the most aggressive animal occupies the most desirable territory. The object of this exercise is to examine the role of dominance in the social organization of captive crayfish.

METHODS

Subjects and Materials

Mature crayfish, *Cambarus*, are available in nature or from biological supply houses. About 20–25 crayfish (at least half males) should be obtained at least two weeks before the planned laboratory exercise. If laboratory space is limited, establish four or five 10-gallon aquaria with gravel on the bottom and filled with water to the top. In addition, provide each tank with an air stone or a Dynaflo Filter®, two or more small clay flower pots (crayfish size) lying on their sides, and a few rocks about 12 cm in diameter. The flower pots and spaces between the rocks create suitable protected habitats for individual animals. Tops of aquaria should be covered with a hood so the crayfish cannot escape by climbing the air tubing or siphons. If there is adequate laboratory space, the 10-gallon aquaria can be replaced by plastic wading pools approximately 2 m in diameter. These should be filled with water to a depth of 5–8 cm and the bottom covered with gravel and a few larger rocks. Flower pots should be provided for shelter, some incorporated in rock piles and others isolated at various locations in the pool (Fig. 29.1). Place five crayfish, including at least two males, in each tank or pool at least two weeks (preferably longer) before testing. Crayfish may be sexed by viewing their ventral surfaces (Fig. 29.2). The first two pairs of swimmerets in males are composed of a protopodite and endopodite fused together, which, when pressed against the opposite modified swimmeret, form a tube used for transfer of sperm to the female. The first pair of swimmerets on females is reduced and the second pair is similar to the others. Feed them small pieces of cooked chicken or earthworms,

Figure 29-1. Top view of prepared wading pool. The bottom is covered with gravel, with flower pots and rocks serving as shelter for the crayfish. Water in pool should be 5 to 8 centimeters deep.

but avoid left-over decaying meat, which will pollute the water: this is especially necessary in the pools unless some kind of filter is provided. Be sure each animal is given an opportunity to feed by placing small bits of food near its shelter. Do not feed the crayfish for three days before the scheduled laboratory exercise.

On the day of testing the class should be divided into as many teams as there are tanks, and each team should be equipped with an empty 10-gallon aquarium or pan of equal floor area (capable of holding 5–8 cm of water), five clean coffee cans, a small flower pot, and small pieces of fresh fish (soluble material in fish attracts crayfish faster than chicken or earthworms).

Procedure

A. Are Crayfish Territorial? As a first step, each student team should unobtrusively record the location and identifying marks of each undisturbed crayfish occupying a sheltered spot in a holding tank. Record this information on a sketch of the physical layout of the tank. Then capture one of the five crayfish in the tank and move it to the entrance of its nearest neighbor's shelter. You can capture a resting animal by moving your hand toward it from the rear, grasping the carapace with thumb and forefinger while pressing the animal against the substrate. When you have a firm grip the animal may be lifted. Does the crayfish attempt to defend itself against your hand? List the evasive or agonistic-defensive measures utilized by the animal. Release the crayfish in front of its neighbor's shelter and observe the behavior of intruder and resident for three minutes. Record your observations and repeat the procedure for the other crayfish, always moving a captured crayfish to the entrance of an undisturbed neighbor's shelter. Does the behavior you have observed in displaced and resident crayfish provide evidence for a dominance hierarchy (dominance is independent of location of conspecific) or territoriality (each animal is dominant over all others in its own territory and subordinate to all others in their territories)? Do all specimens show the same general behavior patterns?

B. Evidence of an Established Dominance Hierarchy Capture the five crayfish in the holding tank assigned to your team. Isolate each in a clean coffee can filled with water to a depth of 2 cm. Drain enough water from the holding tank to fill your empty aquarium or plastic pan to a depth of 5–8 cm. Lower the laboratory light intensity if possible. Introduce the five crayfish into the tank one at a time at two to five minute intervals. Have each member of your group observe the behavior of one crayfish. Characterize encounters between crayfish in the tank, noting which animal is dominant and subordinate for

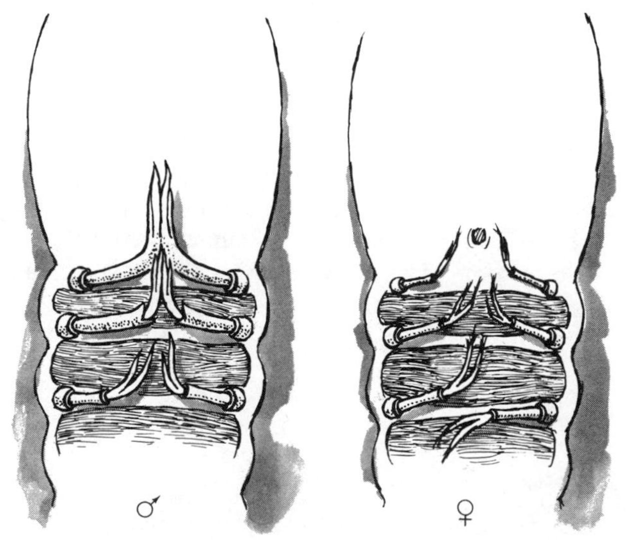

Figure 29-2. Ventral view of crayfish showing sex differences.

each encounter. Gently prod the animals if they do not move about the tank freely. Describe a typical encounter between two animals. Do crayfish direct their large chelae (pincers) at any particular part of the opponent's anatomy? Are chelae open or closed in encounters?

After 30 minutes of observation, are you able to determine the rank of each crayfish in a dominance hierarchy? Place a small piece of chicken, earthworm, or fish close to the animal adjudged to be the lowest in the hierarchy. What happens when a dominant animal approaches? How do different pairs of appendages function in prey capture and ingestive behavior? Add a flower pot or can to the tank. Then gently prod individuals into passing its entrance. Is it permanently occupied or are individuals displaced depending on their rank in the hierarchy? Do the crayfish show wall-seeking behavior (see Exercise 13)?

Sex and weigh each of the five animals and relate your findings to hierarchical rank.

QUESTIONS

1. Based on your data, what might happen if an aggressive crayfish established a territory consisting of the entire floor area of an aquarium and a subordinate specimen was forced to remain in this territory?

2. Would it be better to place five individuals in a holding tank simultaneously or separately over a period of two or three days? Why?

3. Would a rank change occur in an established hierarchy if you removed one or both large chelae from an individual high in rank? Write a hypothesis and outline a test procedure for the hypothesis.

4. At times molting crayfish in holding tanks are killed by conspecifics. Can you suggest a reason for this?

5. In what ways is territoriality adaptive for a species? Can you suggest any advantages of the dominance hierarchy in a nonterritorial species?

ADDITIONAL STUDIES

You can study the formation of hierarchies by matching dominant animals from different holding tanks in a neutral arena. How are the encounters different from those observed previously? Mapping of territories and effects of competition can be studied in a wading pool by varying the number and characteristics of shelters. How are territorial behavior and other social interactions influenced by molting and mating?

REFERENCES

Bruno, M. S., 1968. *Crayfish.* New York: McGraw-Hill.

Eibl-Eibesfeldt, I., 1961. The fighting behavior of animals. *Sci. Amer.* 205(6):112–122. (Offprint 470.)

Lowe, M. E., 1956. Dominance-subordinance relationships in *Cambarellus shufeldtii. Tulane Studies Zool.* 4:139–170.

Tinbergen, N., 1968. *Social behaviour in animals.* London: Methuen.

Waterman, T. H. (Ed.), 1961. *The physiology of the crustacea.* Vol. II. *Sense organs, integration, and behavior.* New York: Academic Press.

JOSEPH W. JENNINGS DONALD L. OLSON
University of Montana

Releasers for Agonistic Display in Male Siamese Fighting Fish

The concept of social releasers is particularly important in our understanding of the innate behavioral patterns of animals. For example, the male Siamese fighting fish, *Betta splendens,* will exhibit its full agonistic display only under certain stimulus conditions (e.g., when another male *Betta* intrudes into its defended area). However, a male *Betta* is a complex of stimuli, including outline, color, brightness, and pattern of movements. Your task in this exercise is to discover the necessary and sufficient stimuli that constitute the releasers for the agonistic displays of male *Bettas.*

The species-specific agonistic displays released by key stimuli of another male *Betta* communicate the hostile intent to the rival in a ritualized series of responses and counter responses. Such displays permit a dominant-subordinate relationship to be established with minimum physical harm to either fish. Thus, in the field, either the intruder or the resident fish may flee after a ritual exchange of sign stimuli.

Simpson (1968) has made a careful and detailed study of the behavior of this species. Figure 30.1 shows its typical agonistic displays.

METHODS

Subjects and Materials

Male Siamese fighting fish show a great deal of variability in aggressiveness. Since the more aggressive fish will give you the best results, conduct some preliminary tests to determine the differences in aggressiveness between individuals. Place a mirror against the side of the holding container or aquarium of each fish. The more aggressive fish should promptly and fully display in response to their own reflection in the mirror.

Bettas should be individually stored in containers which provide both physical and visual isolation. Maintain a steady water temperature as close to 25–27°C as possible to get the best results from your fish. Consult a pet shop or library for details on the care and feeding of *Bettas.*

Each team of students will need a 1- or 2-liter flat-sided goldfish bowl. Also needed are a mirror large enough to cover most of one side of the bowl, a pack of blank white file cards about 10 by 15 cm, and crayons or colored pencils or felt-tip pens including a fairly clear red, yellow, blue, and green. Assemble all materials before beginning the first step.

Procedure

Set the mirror flush against the side of the bowl. Observe the displays made by the fish toward its reflection for about five minutes. Write a description of the fish and its behavior at this point and revise it as you continue the exercise.

Remove the mirror and attempt to draw a picture of a male *Betta* on a file card from what you have observed and described, without looking directly at the fish in the bowl. Attach your drawing to the side of the tank with tape. If the fish behaves toward your drawing as it did toward its own reflection, then your picture has included the necessary and sufficient releasers. If the fish does not react fully to your drawing, at least some of the releasers have eluded you, either in your description or in its translation into a drawing. Begin again with the mirror.

Even if your description and drawing have proved adequate, you have not really discovered the releasers, you have only shown that you were able to reconstitute something that qualifies as a male *Betta* to both you and your fish. Now begin abstracting your drawing along every possible physical dimension that you first used to describe your fish. This can be accomplished in either of two ways. You can begin by carefully eliminating details from your drawing until the fish stops responding or you may start by abstracting one physical dimension—such as size, shape, or color—and adding one dimension at a time until the fish responds. If you are systematic, you should discover that certain combinations of color, size, and shape produce a stronger response from your fish than other combinations of the separate components. This effect has been called heterogeneous summation. In either case, carefully record the step-by-step changes in your drawings and save each drawing for future reference. No matter which of the two approaches you try first, both should eventually be used. When you have completed this task the final drawings produced by both procedures should be similar if the releasers have been isolated. What you have discovered are the basic visual stimuli which act as a signal that another male *Betta* is close by.

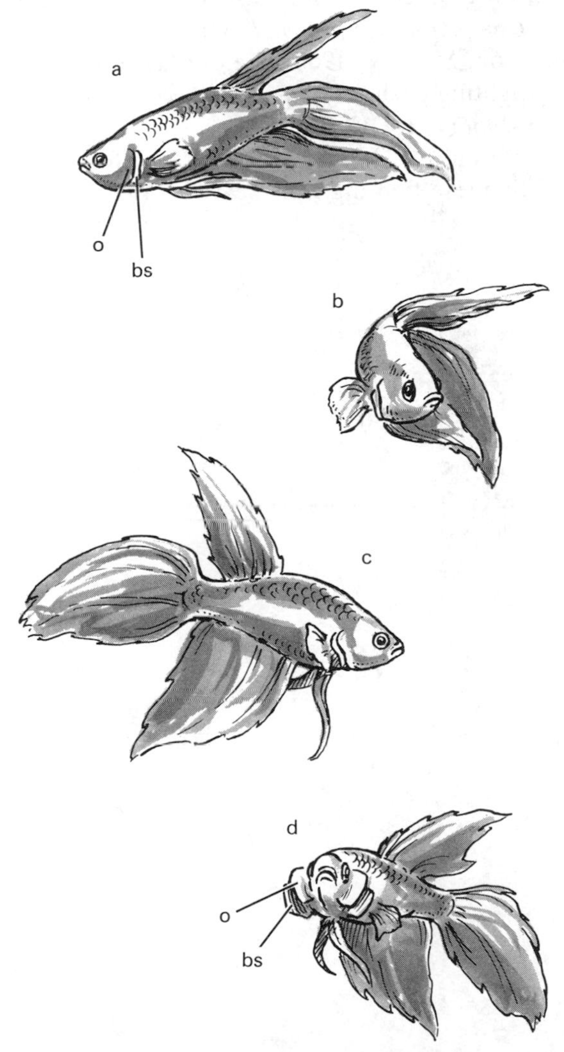

Figure 30-1. Agonistic display patterns of male Siamese fighting fish (*a* and *b*: nondisplaying; *c* and *d*: displaying). In *a*, the fins, as they are placed round the fish in a clockwise direction, are the dorsal, caudal, anal (also called ventral), paired pelvic fins, and one of the paired pectoral fins; *bs* = black branchiostegal membrane; *o* = operculum. In *c*, the fish is broadside, with the pelvic fin nearest to its partner erect. In *d*, the fish has its gill covers erect and is facing a stationary puppet. Its pelvic fins are partly withdrawn. (After Simpson, 1968.)

DISCUSSION

The display of the *Betta* is basically innate. However, experience can influence the fish's behavior. Thompson (1963) and Hogan (1967) demonstrated that the release of the aggressive display can serve as a positive reinforcement for the male *Betta* in learning a new behavior (such as swimming through a ring) in order to be re-exposed to the releaser. An illustration of Thompson's apparatus can be found in Hinde (1970, p. 345).

Long periods of repeated exposure of a *Betta* to its mirror image or model may lead to habituation (defined in Exercise 9) of the usual response (Baenninger, 1970). Eventually, this cessation of responding may develop into an aversion for its image or the model. Thus, repeated and environmentally inappropriate conditions lead to modifications of innate behavior that range from a lack of responsiveness to a negative or aversive reaction.

QUESTION

Speculate why the male *Betta* exhibits elaborate agonistic displays towards an intruder rather than immediately engaging in physical combat. What other animals go through complex agonistic displays before damaging or fatal fights take place?

ADDITIONAL STUDY

Hess (1953) has reported that temperature can influence the aggressiveness of male *Bettas,* thus giving us an idea of the pitfalls in the stimulus-response approach to animal behavior. The relationship between releaser and the behavior released is not a simple mechanical one. Like all evolutionarily adapted systems, the context in which the stimuli are presented is all-important. For an additional study design an experiment to test the effect of temperature on the frequency and duration of agonistic display in this species.

REFERENCES

Baenninger, R., 1970. Visual reinforcement, habituation, and prior social experience of Siamese fighting fish. *J. Comp. Physiol. Psychol.* 71:1–5.

Hess, E., 1953. Temperature as a regulation of the attack response of *Betta splendens. Z. Tierpsychol.* 9:379–382.

Hinde, R. A., 1970. *Animal behavior.* 2d ed. New York: McGraw-Hill.

Hogan, J. A., 1967. Fighting and reinforcement in the Siamese fighting fish (*Betta splendens*). *J. Comp. Physiol. Psychol.* 64:356–359.

Simpson, M. J. A., 1968. The display of the Siamese fighting fish, *Betta splendens. Anim. Behav. Monogr.* 1(1):1–73.

Thompson, T. I., 1963. Visual reinforcement in Siamese fighting fish. *Science* 141:55–57.

MILES H. A. KEENLEYSIDE
University of Western Ontario

31

Schooling Behavior in Fish

Fish swimming together in a school exhibit a conspicuous form of social organization. Typically, all individuals in a school are the same species and size, there is no one "leader," and all of the fish are usually engaged in the same activity at the same time. Many other animals (e.g., some birds, mammals, and invertebrates) show a similar type of mass grouping behavior during at least part of their lives, but fish are among the best examples for use in a laboratory study of the phenomenon. This is because many small freshwater fishes school consistently, even in the unnatural surroundings of a 15-gallon aquarium.

A fish school has been defined as a group in which the animals stay together because they are showing positive social responses to each other, not because they are responding similarly to a common external factor. Thus, groupings that are formed when many fish approach a localized external stimulus (such as a spotlight or a concentration of food), but that break up when the external stimulus is diffuse, are not considered true schools. Your objective in this exercise is to gather evidence which may help to explain the uniformity in species, size, and behavior of fish in a school.

METHODS

Subjects and Materials

Some species of fish suitable for the experiments in this exercise are the following:

Zebra fish (*Brachydanio rerio*)
Harlequin fish (*Rasbora heteromorpha*)
Scissor-tail (*Rasbora trilineata*)
Rosy tetra (*Hyphessobrycon rosaceus*)
Tiger barb (*Barbus tetrazona*)
Pristella (*Pristella riddlei*)
Brook stickleback (*Culaea inconstans*)
Juvenile convict cichlids (*Cichlasoma nigrofasciatum*)
Juvenile Jack Dempseys (*Cichlasoma octofasciatum*)

Each team of students will require an aquarium (approximately 15-gallon size) and two glass jars or transparent containers (see Fig. 31.1). A stopwatch and other recording materials will also be needed.

Procedure

A. Preliminary Observations First, if a large community aquarium has been set up in your laboratory, study the grouping tendencies of the fish species in it.

1. Do they all swim in single-species groups? Are there any signs of intra-group aggressive behavior?
2. Does the distance between the fish in a group appear to change as the group moves about the aquarium?

3. What effects on grouping behavior result from tapping on the aquarium frame and from introducing food?

After making these observations select two species and perform the following two experiments in your own aquarium. If time is short, half the class can conduct one experiment and half the other.

B. Species Affinity In selecting the fish for this experiment, choose two species of about equal size. Some students should choose species that appear to swim together, at least occasionally, in the community aquarium; others should choose species that do not swim together.

Draw two vertical lines on the front wall of your aquarium, dividing it into three equal areas. Place equal numbers of the two species of fish in separate screw-top jars, one species per jar. The number of fish will depend on their size; eight to ten 3- to 5-cm long fish in a 4-liter jar is adequate (Fig. 31.1). Place one jar containing a group of fish at each end of the aquarium. Gently place a single fish of one of the two species (call it the Test Fish) in the center of the aquarium. Release the Test Fish carefully, making sure you do not direct it toward one of the jars at the start of the test. One technique to reduce bias here is to place the Test Fish inside an inverted glass jar in the center of the aquarium for a standard time period (1 to 5 minutes) and then to release it by carefully raising the jar. During the ensuing 15-minute period record the following data in Table 31.1: (*a*) time (in minutes) the Test Fish spends in each of the three areas and (*b*) the number of times the Test Fish moves from one area to another.

Remove the Test Fish, reverse the positions of the jars, replace the Test Fish, and repeat the recording. Remove the Test Fish, use one of the other species as a new Test Fish, and repeat the entire procedure. During the four tests, take notes on apparent interactions between the Test Fish and the fish in the two jars. For example, does the Test Fish swim close to a jar, closely following the movements of a fish inside? Does it turn and quickly swim back to a jar after moving a short distance away? Does sudden activity by the fish in a jar appear to stimulate the Test Fish to swim towards them?

Those using the same two species should combine results at the end of the period.

C. Effect of Group Size on Affinity for Conspecifics Select the species that showed the strongest schooling tendencies in Part B. Place the two jars in opposite ends of the aquarium. In one jar put two fish and in the other six fish, all of the same species (Fig. 31.2). Carefully release a conspecific fish in the center of

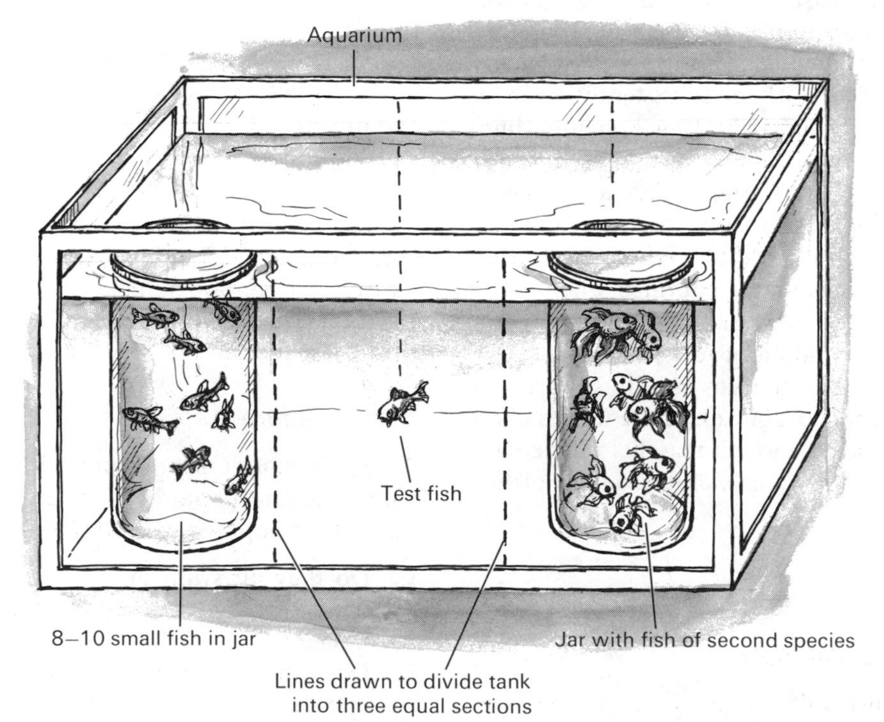

Figure 31-1. Aquarium set-up for studying the tendency of a single fish to school with conspecifics.

Table 31-1. Sample data sheet for recording the tendency of a fish to school with conspecifics.

Test	*Minutes spent in area with*			*Number of moves to new area*
	Conspecifics	*No fish*	*Heterospecifics*	
Test 1				
Test 2				
Means				
Test 3				
Test 4				
Means				

the aquarium. During the ensuing 15 minutes record in Table 31.2 the same data collected in Part B.

Remove the Test Fish, reverse the positions of the jars, replace the Test Fish, and repeat the recording. Place equal numbers of fish in each jar, use a different individual as the Test Fish, and repeat the entire procedure. Again, make qualitative records of the responses shown by the Test Fish and the captive fish towards each other.

Students using the same species of fish should combine their results. The chi-square test (Sokal and Rohlf, 1969) may be used to compare the observed distribution of Test Fish in the aquarium with a theoretical distribution, such as equal time spent in all three areas of the aquarium.

QUESTIONS

1. Did the Test Fish spend more time with its own species than with the other species?
2. Did the Test Fish spend more time with the large or with the small school?
3. In what other ways did the Test Fish react to the fish in the jars?
4. How did the enclosed fish react to the Test Fish?
5. Do you think that the tendency of a Test Fish to stay near the fish in a jar is strong evidence for schooling behavior? Justify your answer.
6. Compare your findings with those in the references.

Figure 31-2. Aquarium set-up for studying the effect of group size on the response of an isolated fish.

Table 31-2. Sample data sheet for recording the effect of group size on the response of an isolated fish.

Test	*Minutes spent in area with*		*Number of moves to new area*
	2 fish	*6 fish*	
Test 1			
Test 2			
Means			
	6 fish	6 fish	
Test 3			
Test 4			
Means			

ADDITIONAL STUDIES

If the Test Fish does show a strong affinity for conspecifics, the next question to answer might be: What are the cues on which this apparent species recognition are based? This may be difficult to answer, but consider the following:

Are visual cues necessary for schooling? Study your species under dim illumination and after the fish have been in complete darkness.

If vision is necessary to maintain schooling, what types of visual cues seem most important? Is movement of other fish in a group attractive? Is regular smooth swimming as attractive as sudden darting and frequent turning? Do patches of conspicuous colors on the body or fins seem to be social releasers for schooling behavior?

How important is size of fish in the maintenance of school cohesion? Will an isolated fish approach and try to school with conspecifics rather than fish of another species even if the latter are closer to its own size?

REFERENCES

Breder, C. M., Jr., 1959. Studies on social groupings in fishes. *Bull. Amer. Mus. Nat. Hist.* 117: 397–481.

Breder, C. M., Jr., 1967. On the survival value of fish schools. *Zoologica* 52:25–40.

Keenleyside, M. H. A., 1955. Some aspects of the schooling behaviour of fish. *Behaviour* 8: 183–248.

Shaw, E., 1962. The schooling of fishes. *Sci. Amer.* 202(6):128–138. (Offprint 124.)

Shaw, E., 1970. Schooling in fishes: Critique and review. *In* L. Aronson, E. Tobach, D. Lehrman, and J. Rosenblatt (Eds.), *Development and evolution of behavior,* pp. 452–480. San Francisco: W. H. Freeman and Company.

Sokal, R. R., and F. J. Rohlf, 1969. *Biometry.* San Francisco: W. H. Freeman and Company.

Steven, D. M., 1959. Studies on the shoaling behavior of fish. I. Responses of two species to changes of illumination and to olfactory stimuli. *J. Exp. Biol.* 36:261–280.

Williams, G. C., 1964. The measurement of consociation among fishes, and comments on the evolution of schooling. *Publ. Mus. Mich. State Univ., Biol. Series* 2:349–384.

BAYARD H. BRATTSTROM
California State University, Fullerton

32

Population Density and Habitat Influences on Social Organization

Two basic forms of social organization are recognized in animal populations—territoriality and the dominance hierarchy (see also Exercises 27, 29, 33 and 35). Aggression, which acts to establish and maintain social organization, arises out of competition for limited resources (Banks and Wilson, 1974; Brown, 1964; Brown and Orians, 1970). Population density, habitat structure, and the distribution of resources have important effects on the frequency and intensity of aggressive responses.

In territoriality an animal defends a specific area (though the area may change shape, move, or be defended only at certain times of the day or year) against others (often just against males) of the same species. Thus a territorial animal has almost exclusive use of an area for a period of time. In the dominance-hierarchy form of social organization there is a rank order of individuals with a most dominant, or alpha, individual at the top of the rank order and other individuals increasingly subordinate to those above them in the rank order. The hierarchy usually develops after much real or ritualized aggression when the animals first come toegther. The frequency of aggression decreases after a few days and the dominance order is maintained by occasional aggressive behaviors or displays. In fish and reptiles the order is usually related to size and color, and thus the alpha male is likely to be the largest and most brightly colored individual. Hierarchies may be linear, or dendritic (with several individuals at various levels of social status), or triangular (see below). The upper levels of a dominance order are usually stable and clearly defined; the lower levels are often unstable and poorly defined.

Territorial social organization is commonly found in animals when the habitat is fairly uniform, productive, and stable, and the population is small in relation to resources. A hierarchy often develops when food or other resources become limited or when animals are crowded (have reached a high density). The transition from territoriality to a hierarchical social organization under adverse conditions allows more animals to live in a given area. It also allows animals to concentrate around a localized but highly concentrated resource (food, water, basking or sleeping site).

The objective of this exercise is to observe the changes in social organization of an animal population (from territoriality to an hierarchical social organization and vice versa) resulting from changes

Table 32-1. Some possible dominance relationships between individuals in the hierarchical form of social organization.

Linear	*Dendritic*	*Triangular*
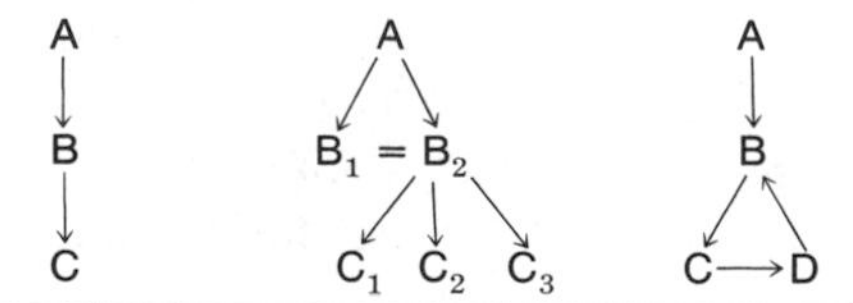 A ↓ B ↓ C	A ↙ ↘ $B_1 = B_2$ ↙ ↓ ↘ C_1 C_2 C_3	A ↓ B ↙ ↖ C → D

in population density and various manipulations of the physical environment.

METHODS

Subjects and Materials

The exercise is designed for use with fish and reptiles. However, it can be easily modified for use with birds, rodents, crayfish, or any normally territorial species. The more territorial tropical fish (e.g., paradise fish, cichlids, etc.) are best: mollies, platys, and swordtails are less territorial. Sunfish and bass locally caught can also be used. Almost all diurnal basking lizards (e.g., such iguanid lizards as fence lizards, *Sceloporus*; desert iguanas, *Dipsosaurus*; side-blotched lizards, *Uta*; or anoles, *Anolis*) and such nocturnal lizards as geckos (*Phyllodactylus* and others) can be used. Lizards can be caught locally or purchased from pet stores or biological supply houses, and can be marked by painting numbers or markings on their back with a felt pen, nail polish, or paint. Fish can be marked by sewing monofilament and beads through their chins or dorsal fins. Studies should be conducted with males only. Nocturnal animals can be studied under red light. Diurnal basking lizards will need a heat source (150-watt incandescent or 150–250-watt infrared "sun lamp") at least during the day. Animals should be regularly fed following observations: lizards do well on meal worms, which are available at most pet stores.

Aquaria of various sizes (20 gallons or larger) can be used. Fish aquaria should contain no objects other than those suggested in the experiments below. Lizard cages can be constructed out of cardboard boxes; the corrugated-paper boxes in which refrigerators are packed make excellent cages.

Procedure

Before you conduct the following experiments become familiar with the aggressive behavior (e.g., postures, displays, etc.) of the test species. Hence, initially observe the test subjects and establish data-recording techniques.

A. Changes in Density Place about ten small territorial fish in the largest aquarium available. After an hour, or on the following day, observe and draw diagrams of the territories of each fish (areas occupied exclusively or defended against others). Count the number of aggressive interactions for each fish for a 10- or 20-minute period. On subsequent days insert a board or screen into the aquarium so that its volume is reduced by one-half, three-fourths, and more. At what density does the social system switch to an hierarchical one? Are all fish included in the hierarchy? Is the same shift in social organization observed if you increase the density in the original aquarium by adding more fish? Is the result the same if you lower the water level? Can the system be switched back to a territorial one, even at high densities, by adding localization objects (e.g., rocks, floating plastic plants, etc.)?

The same experiment can be done with diurnal territorial lizards. If you have a 100-gallon aquarium or a large refrigerator-sized cardboard box, introduce 4 to 6 male anoles or side-blotched lizards or 3 or 4 male lizards if larger species are used. Crowd the lizards on subsequent days by decreasing the available space or by adding new subjects as described above for fish.

B. Habitat Manipulation Place several bricks, vials, flower pots, or rock piles in a 10- or 20-gallon aquarium (Fig. 32.1,*a–d*). Place as many male lizards in the aquarium as there are hiding places. The best lizards to use are nocturnal geckos and skinks. Geckos with toe pads do well with bricks and crevices; ground geckos and skinks do better with vials or jar lids set at an angle. After the animals have established territories within each of the cracks or other hiding places, systematically remove one brick, vial, etc. at a time (e.g., each day). As the number of cracks or hiding places is reduced, evicted animals will attempt to gain entrance to a crack occupied by another male. How do the social organization and dominant-subordinate relationships change as the number of hiding places is recorded?

Figure 32.1,*a* shows a simple cage set up for observing territorial behavior in small diurnal basking reptiles such as fence lizards or side-blotched lizards. With all of the heat lamps on, each of four males placed in such an aquarium will normally establish a territory on a rock. Since these lizards must obtain their body heat from the external environment (the lamps), the social system can be altered either by reducing the number of rocks or rock piles (Fig. 32.1,*e*) or by turning off all but one of the heat lamps. Experiment with other set-ups. Try switching the end lamps alternately on and off. Is the same male always

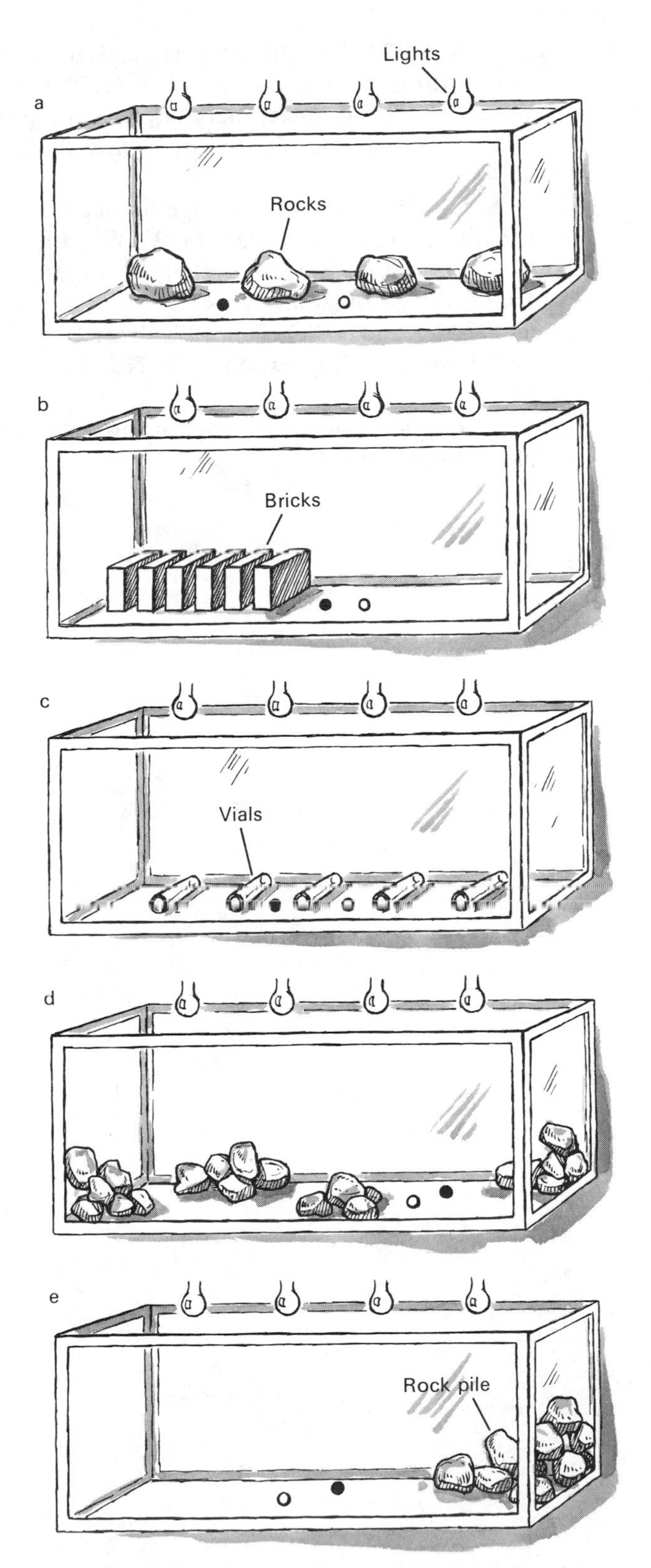

Figure 32-1. Typical laboratory set-ups for studying changes in social organization of small populations of lizards. Small circles are possible places to put food and water.

dominant or is the previous holder of the heated rock dominant?

Most of the experiments described can be conducted within a 3–5 day period with 10 to 30 minutes of observation per day. Some fish and fence lizards will react quickly to the manipulations suggested and thus some changes may be observed within a 2- or 3-hour laboratory period, especially if aquaria or enclosures are set up earlier.

QUESTIONS

1. How does the relative size and color of the males affect social status?
2. What are some of the factors that may determine whether a hierarchy is linear, dendritic, or triangular?
3. Why are some hierarchies stable whereas others are not?
4. Can the results of these studies be applied to problems of human overpopulation and crowding? To the design of cities or classrooms?

ADDITIONAL STUDIES

The above experiments are relatively open-ended since a host of alternative manipulations may easily be applied. What happens when the alpha individual is removed? Does he regain his top-ranking position when reinstated in the group? Does the length of time away from the group influence the capacity of an individual to regain its former dominance status?

What happens to the social status of an individual lizard when its bright colors are covered with paint? What experiments with models could be devised?

The principles of behavior demonstrated in this exercise could also be studied in a natural environment. Find a population of territorial lizards or fish. Map the territories of the males. Try watching territorial bass or sunfish with a face-plate and snorkel; write under water with crayons on heavy plastic sheets. Is the territory used or defended all the time? How is the territory defended? Do different-sized males have different-sized territories? If possible, collect and mark the animals observed and bring them into the laboratory. Using these animals conduct one or more of the previously described experiments.

REFERENCES

Banks, E., and M. Wilson (Eds.), 1974. The ecology and evolution of social organization. *Amer. Zool.* 14:7–264.

Brattstrom, B. H., 1971. Social and thermoregulatory behavior of the Bearded Dragon, *Amphibolurus barbatus. Copeia* 1971:484–497.

Brown, J., 1964. The evolution of diversity in avian territorial systems. *Wilson Bull.* 76:160–169.

Brown, J., and G. Orians, 1970. Spacing patterns in mobile animals. *Annu. Rev. Ecol. Syst.* 1: 239–262.

Calhoun, J. B., 1962. Population density and social pathology. *Sci. Amer.* 206(2):139–146. (Offprint 506.)

Desor, J. A., 1972. Towards a psychological theory of crowding. *J. Pers. Soc. Psychol.* 21:79–83.

Erickson, J. B., 1967. Social hierarchy, territoriality, and stress reactions in sunfish. *Physiol. Zool.* 40:40–48.

Rand, A. S., 1967. The adaptive significance of territoriality in Iguanid lizards. *In* W. W. Milstead (Ed.), *Lizard ecology: A symposium.* Columbia: University of Missouri Press.

Scheflen, A. E., and A. Scheflen, 1972. *Body language and social order.* Englewood Cliffs, N.J.: Prentice-Hall.

Sommer, R., 1969. *Personal space.* Englewood Cliffs, N.J.: Prentice-Hall.

ALLEN W. STOKES
Utah State University

33

Social Organization and Courtship Behavior in Chickens

Chickens, like many other animals, develop a social hierarchy after being placed together for some time. In this exercise you will observe the behavior of a small group of cockerels when they are first placed together as strangers in a pen, and then compare the behavior of this "disorganized" group with that of a similar group that has been living together for several weeks. In this way you can observe the type of social hierarchy that develops in chickens, the behavioral means by which this organization develops, and the changes in behavior in the group once organization has occurred. Finally, you will observe courtship and sexual behavior by placing a hen with the cockerels.

METHODS

Subjects and Materials

The exercise requires ten cockerels and one hen of a white breed of chicken; the more fully grown they are the better. Mark each of the birds individually on the head or back with felt pens of different colors for ready recognition (if a white breed is not available, use colored leg bands). Four cockerels should be placed together in one of two observation pens for a week or more prior to observation. The remaining six should be placed in separate cages out of sight of each other. The two observation pens should be 3 meters square or larger and screened so that the birds in the two pens are visually isolated. It is best to have no escape shelter or roosts inside the pens during observation periods. A poultryman's standard hook is helpful for catching a bird by the leg when removing it from a pen (Fig. 33.1).

Procedure

A. Behavior of Strangers Once strange cockerels are placed in a neutral pen the interactions are likely to begin quickly and be intense. Therefore, first place

Figure 33-1. Use of poultryman's hook in capturing chickens.

just two of the isolated cockerels together as a pilot study. Each student should be assigned one or the other of the birds and should be responsible for observing and recording data on only that bird. A sample form for recording data in shorthand is suggested below:

Bird: *Red*

Encounter with	*Behavior by Red*	*Winner*
1. Blue	Ap;Wf;Pk;Lat;Wz;Cir;Lat;Wz; Pg;Av;	
2. etc.	etc.	
3. etc.	etc.	

Key:
Ap —approach rival
Wf —wing flap
Pg —peck at ground
Pk —peck rival
Lat—lateral orientation to rival
Cir—circling rival
Wz—waltz (circling with outer wing dropped)
Av —avoiding rival

When the members of the class feel that they have had enough experience in recording data, remove the two cockerels, and place the four other isolated cockerels in the pen. Proceed as before, with each person responsible for recording the behavior of only one bird. Observe for at least six 5-minute periods—as many as ten if the action persists. Keep your observations for each 5-minute period separate so that you can quantify changes in frequency and types of behavior for the entire period of observation.

Pool all observations and arrange the data for wins and losses in the form suggested by Table 33.1.

1. On the basis of pecks and fights can you assign a social ranking for each individual in the group?

2. Are some behavior patterns associated with dominance and others with subordination? Are the differences between these categories clearly defined or are the categories on a continuum?

3. Did the frequency or intensity of certain behaviors change over successive 5-minute periods of data recording?

B. Behavior of Organized Flock Now observe the group of cockerels that has been together in one observation pen for at least one week. The birds should have been deprived of food for 36 hours before beginning observations. Place a food pan in the pen, and cover the top so that only one bird can feed at a time. Again, each student should record the behavior of a single bird on the check list. A 20-minute period should suffice to determine the social ranking of these birds.

1. What criteria proved most helpful in determining the social rank—the overt aggressive acts, such as pecking, or the associated feather and body movements?

2. How would you characterize the differences between an organized and disorganized chicken flock?

3. Was there any correlation between social rank and size of the birds? Between social rank and comb size?

4. Did the frequency or intensity of certain behaviors change in the course of the observation period? How would you categorize those patterns that declined in frequency during this period?

C. Courtship Behavior Now place a strange hen in with the group of four cockerels constituting the organized flock. Each student should be assigned

Table 33-1. Social ranking of four cockerels in the same pen based on outcome of social encounters.

		Losses by bird				*Total wins for each*
		1	2	3	4	
Wins by bird*	1	—				
	2		—			
	3			—		
	4				—	
Total losses for each						

*A "win" may be a peck at rival, an actual win in a fight, or a threat by winner associated with avoidance by loser. Birds are arranged in descending order of dominance.

one of the birds—a cockerel or the hen—and should record on a form sheet the behavior patterns of only that bird. Be prepared for instant action in order not to miss the initial responses of both the cockerels and the hen. It may be necessary to add to the list of behaviors some behavior patterns that have not previously been observed. Record the behavior for 10 or 15 minutes; then remove the hen.

1. What behavior patterns did these cockerels show toward the hen that they also showed toward other cockerels?
2. What behavior patterns seem restricted to heterosexual interaction?
3. What are the essentials of courtship?
4. Does courtship share some of the same behavior patterns as agonistic behavior? Why should this be? Davis (1964) will be helpful in interpreting your observations.
5. Does the response of an individual cockerel to the hen have any relation to his position in the social hierarchy?
6. Is social dominance in chickens advantageous in securing a mate?

ADDITIONAL STUDIES

The following questions are designed to stimulate further experimentation. Consult the references at the end of the exercise for further information on these topics.

1. Would the behavior of (*a*) the hen and (*b*) the cockerels be different depending on whether they were sexually experienced or sexually naive, i.e., if they had never seen a bird of the opposite sex before being placed together?
2. If testosterone is injected into a cockerel, does it change his social rank?
3. How would the behavior of a hen injected with testosterone be affected? Why?
4. If the hen and cockerels were left together for a week, what would be their response to a second hen placed in with them? Are the two hens compatible? Does the dominant male court the second hen and form a harem, as jungle fowl do in the wild, or do lower-ranking males court this second hen?
5. Would a flock of hens also establish a social hierarchy? How would dominance be expressed?
6. How does starvation affect levels of aggression when food is restored?
7. How would cockerels react to younger birds of the same sex?
8. If one bird is materially altered in appearance (e.g., through extensive coloring of its plumage), how does this treatment affect its acceptance by its penmates?

REFERENCES

Banks, E. M., 1960. *Social organization in the red jungle fowl* (Film No. PCR-119K; 10-minute, 16-mm, silent, color). University Park, Pa.: Psychological Cinema Register, Pennsylvania State University.

Davis, D. E., 1964. The physiological analysis of aggressive behavior. *In* W. Etkin (Ed.), *Social behavior and organization among vertebrates,* pp. 53–74. Chicago: University of Chicago Press.

Guhl, A. M., 1956. The social order of chickens. *Sci. Amer.* 194(2):42–46. (Offprint 471.)

Guhl, A. M., 1962. The behaviour of chickens. *In* E. S. E. Hafez (Ed.), *The behaviour of domestic animals,* pp. 491–530. Baltimore: Williams & Wilkins.

Guhl, A. M. *Social behavior in chickens* (A BSCS Single Topic Inquiry Film; 4-minute, super 8-mm film loop). Boulder, Colorado: Biological Sciences Curriculum Study.

THREE

FIELD STUDIES

PETER N. WITT
North Carolina Department of Mental Health
Raleigh, N.C.

Web Construction in Spiders

The orb web of a spider (Fig. 34.1) is the outcome of highly complex and stereotyped behavioral patterns, and has a key role in the survival of the organism. It constitutes a trap in which flying prey is caught, it extends the perceptual range of a nearly blind organism, and the threads provide the substrate on which hooked legs can move at high speed. The purpose of this exercise is to demonstrate some of the factors that determine web location, the sequence of events that results in web construction, and the extent to which web construction may be modified.

Figure 34-1. Web of adult female *Araneus diadematus* Cl. built in an aluminum frame in the laboratory, and photographed against a black box with lights shining from four sides. Note spider as white figure in center of web, and scale of two vertical white lines, 20 mm apart, in upper left corner.

METHODS

Subjects and Materials

For indoor observation you will need a frame to house the spider and provide an anchor for its web, a room in which light and temperature can be manipulated, a good light, and a dark background for observation and possibly photography. A simple frame can be made of four rectangular pieces of cardboard, 10 × 50 cm, taped together to form a square; transparent wrap can be used to close front and back. Figure 34.2 shows a more permanent frame that can be constructed from wood or metal. The screens on the narrow sides of the frame facilitate anchoring for frame threads, and the glass doors prevent escape and permit a clear view of the web; glass should be removed for photography. Web-building indoors is more likely under conditions of (*a*) high humidity, achieved by placing open vessels filled with water around the frame, (*b*) short (8-hour) nights, produced in winter by turning on electric lights in the evening and morning, and (*c*) a temperature difference of at least 5°C between night and day, obtained by changing the thermostat setting or opening windows. More details are given in Witt (1971) and Witt et al. (1968).

Procedure

A. Time and Place of Web Building Since most spiders in the field build their webs every morning at sunrise, observations on the beginning and ending of web-building can be most profitably made at this time. If the exercise takes place in the late summer, you will find many webs built close together, and you can gain some knowledge of a species' site preferences by determining the height of the web from the ground and the direction that it faces as well as by recording the supporting plants. The effect of wind on web site selection can be tested by constructing wind shields or placing plastic cylinders outdoors and watching the consequent web locations, or releasing animals indoors in a space that contains rapidly moving as well as quiet air spaces (Enders, 1972). Additional observations on web height can be made by releasing spiders of different species in a room with a high ceiling. Certain species may build their webs closer to the ceiling than others. (Data on web height can be statistically analyzed for species differences.) For further reading on timing and frequency of web building and on web location see Witt (1963) and Turnbull (1973).

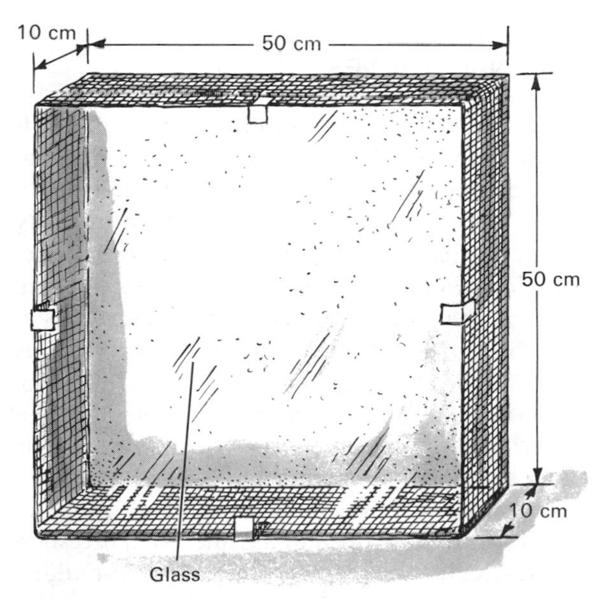

Figure 34-2. Sketch of wooden or aluminum box for keeping a single web-building spider in the laboratory.

B. Web-Building Behavior Since the timing of web building indoors is often unpredictable, the observer may be well rewarded for staying up one night with 3–5 spiders. Best results are obtained by using young females of a species that builds a web daily, e.g., *Araneus diadematus* Cl. (See also Witt, 1971.) Observe the web-building process from beginning to end.

1. How many phases can you recognize during construction of one web?
2. What characteristic pattern of movements is repeated over and over in a phase?
3. Which principal role do the front legs play in web building? Which do the hind-legs play?
4. What is the last movement of the spider when it finishes a web?
5. How much does a spider use its weight in web construction, and how do you think weightlessness would interfere with web building?

Once you have learned the normal sequence of events in web construction try some experimental manipulations as new webs are being built. For example, what is the effect of repeatedly cutting a radius just after it is laid? How many times will the spider rebuild a radius in the same place? Does the spider's response to this experimental manipulation depend on the stage of web construction at the time? What is the consequence of transferring spiders in the middle of construction from their own to another structure?

Through these observations you can learn a great deal about the plasticity of web-building behavior, the way it is "coded" in the animal, and how it serves the survival of the species. (See also Peters, 1970; Reed, 1969.)

C. The Physical Web and Its Behavioral Correlates Only after we have established objective, quantitative measures of behavior can we begin to identify factors that influence the physical structure of the web itself. Simple web measures can be obtained by measuring the vertical and horizontal diameter of the catching area with a ruler and by counting the number of radii and spiral turns; size of catching area divided by number of radii plus number of spiral turns provides a figure for mean mesh size.

For exact measurements, webs should be photographed. You can use pictures taken in the field, but for analysis of pattern, you generally get better photographs if the web is built in a laboratory frame. For further information on web photography see Witt (1971).

A planimeter can be used to measure central, catching, and frame zones of the web separately (Fig. 34.3). These measurements may show some relation to other parameters of importance to the survival of spiders (e.g., abundance of food, size and species of spider). Regularity measures have been shown to be most affected by age and drugs; measures of regularity can be obtained from radial angles as well as spiral distances. If you lay a ruler and protractor over a photograph of a web, centering the protractor on the web center, the central angles of radii can easily be recorded. By subtracting neighboring angles from each other (e.g., 18° − 16° = 2°, 16° − 17° = 1°, . . .) and calculating the mean of all such differences, you obtain a single figure for the angular regularity of a web. Another group of interesting measures identifies the shape (i.e., length over width of catching area). For more explanations of web measures, how they are taken and when they deviate from normal, consult Witt et al. (1968). After obtaining measures for several webs of a single spider, you may want to investigate the role which the absence of a leg plays in the achievement of web geometry. It would be surprising to find that an occurrence as common as the loss of a single leg made spiders incapable of weaving a trap for prey. Legs can substitute in function for each other, and only a multiple loss will result in abnormal web construction. Each pair of legs has a different function (e.g., first pair, probing; last pair, thread laying) and the removal of a pair will show disturbances in web geometry related to that pair's task. New observations may contribute to the analysis of the role which each leg or leg combination plays in thread positioning (see Reed et al., 1965).

Figure 34-3. Thick white lines outline (from the center) the spiral turn enclosing the center area; the outermost spiral turn, enclosing the spiral or catching area; and the frame of the web of an adult female spider. The three areas, which can be measured with a planimeter, show sizes and ratios to each other that are characteristic of the species, age, and state of the builder.

QUESTIONS

1. Is there a species or age difference for: (*a*) mean distance between hub and ground in orb webs? (*b*) vegetation on which webs are attached? (*c*) mean mesh width and size of center area?
2. Is frequency of web renewal dependent on: (*a*) destruction of old web? (*b*) amount of prey eaten? (*c*) species of spider? (*d*) time of removal of old web?
3. List separately parts of web-building behavior or web use which can or cannot be modified by changes in the environment.
4. What factors (biochemical, behavioral) might cause a spider to build a smaller web than normally expected? Design an experiment to determine whether a relatively small web was the result of a lack of silk supply at web-building time.

ADDITIONAL STUDIES

You may want to explore a little further the "world of touch" in which a spider lives. With a vibrating tuning fork held to a radius, questions can be explored like range of perception, threshold intensity, and type of habituation which occurs to repeated stimulation of one or several radii. In prey catching, there is a chain of stimulus-reaction sequences to be observed, from the first impact of the fly against the web to the final sucking out of prey (Robinson and Olazarri, 1971). What stimulus releases what reaction? Are there alternate pathways? Do the spider's age and/or body weight as well as hunger play a role in the hunter-prey interaction? There are many such problems open to the interested investigator with access to a small colony of web-building spiders.

REFERENCES

Enders, F., 1973. Selection of habitat by the spider *Argiope aurantia* Lucas (Araneidae). *Amer. Midland. Natur.* 90:47–55.

Gertsch, W. J., 1964. *American spiders.* New York: Van Nostrand.

Kaston, B. S., and E. Kaston, 1953. *How to know the spiders.* Dubuque, Iowa: W. C. Brown.

Levi, H. W., L. R. Levi, and H. S. Zim, 1968. *A guide to spiders and their kin.* New York: Golden Press.

Peters, P., 1970. Orb web construction: interaction of spider (*Araneus diadematus* Cl.) and thread configuration. *Anim. Behav.* 18:478–484.

Reed, C. F., 1969. Cues in the web-building process. *Amer. Zool.* 9:211–221.

Reed, C. F., P. N. Witt, and R. L. Jones, 1965. The measuring function of the first legs of *Araneus diadematus* Cl. *Behaviour* 24:98–119.

Robinson, M. H., and J. Olazarri, 1971. Units of behavior and complex sequences in the predatory behavior of *Argiope argentata. Smithsonian Contrib. Zool.* 65:1–36.

Turnbull, A. L., 1973. Ecology of the true spiders. *Annu. Rev. Entomol.* 18:305–348.

Witt, P. N., 1973. Environment in relation to behavior of spiders. *Arch. Environ. Health* 7:4–12.

Witt, P. N., 1971. Instructions for working with web-building spiders in the laboratory. *BioScience* 21:23–25.

Witt, P. N., C. F. Reed, and D. B. Peakall, 1968. *A spider's web. Problems in regulatory biology.* New York: Springer.

Witt, P. N., and L. Salzmann, 1973. *Life on a thread, an introduction into the study of behavior* (20-minute, 16-mm film). Philadelphia: Blue Flower Films.

35

BEDFORD M. VESTAL
University of Missouri, St. Louis

Territorial Behavior in Dragonflies

Territorial behavior functions to maintain an area for the exclusive use of one or several individuals. The exclusiveness may be maintained by direct defense of the area against other individuals or by more covert means such as marking of boundaries. Although the territorial behavior of many species has been described, it is most easily studied in conspicuous diurnal forms maintaining territories by overt aggressive behavior. Dragonflies are such a form in which males of some species maintain temporary mating territories along the banks of streams and ponds.

The purpose of this exercise is to observe territorial behavior and certain aspects of the behavioral ecology of dragonflies.

METHODS

Subjects and Materials

Dragonflies may be found during the warmer months along the banks of lakes, ponds, and streams. If you can find several small water sources near each other or a water source with a long bank area you may be able to observe large numbers of these insects. Dragonflies are generally more active on warm, sunny days than on cold and cloudy ones, are more active during mid-day than early or late, and according to one author, become active around 24°C with a peak in activity around 29–31°C. Peak activity times and temperatures may vary with species and geographical area.

Males may be distinguised from females by their positions during mating. The male lands on the back of the female and seizes her head with his anal claspers. As they fly away the female may bend her abdomen ventrally and forward until her genitalia are in contact with the male's genitalia, which are on the ventral side of his second abdominal segment. Pairs may be seen flying either in tandem with the male grasping the female or in the copulatory posture. Males have three terminal abdominal appendages, two above and one below. Females generally have only two dorsal terminal appendages. In some groups the female's ovipositor on the ventral side of the terminal abdominal segment gives the end of the abdomen a somewhat swollen appearance.

Field materials which may be useful are a small pocket notebook, a stopwatch, an insect net (preferably large to avoid wing injury), a tape measure 10 to 50 meters long, binoculars, fine dark cotton thread, a stick or dowel rod several meters long; other items might include a thermometer, a field guide to insects, collecting jars, a hand lens, small containers of various colors of paint (such as Testor's plastic enamel), and insect repellent.

Procedure

Find an area with active dragonflies and space the members of the class along the water's edge. Since dragonflies are rapid fliers and can cover long distances, several observers can more effectively observe individual animals than can one person. Move as little as possible, as dragonflies have good eyesight. Observe the movement of an individual dragonfly. Does the fly seem to follow a regular circuit? Sketch an outline of the body of water, including landmarks, for your orientation. Try to observe the fly for several circuits and record the pattern of his movements on your sketch. Are his movements generally restricted to a given area? If so, he may be confining his activity to his territory. A primary criterion of territoriality is the exclusive use of an area by an individual and possibly a mate. In flying the circuits the male may be patrolling the boundaries of a territory. By watching quietly for some time, you will probably see the fly interact with other animals which enter the territory. How does a male react to other dragonfly males? females? damsel flies? other insects such as butterflies? potential predators such as birds or humans? Is it more important for the owner of a territory to defend its territory against potential competitors of the same species and sex, or against all possible intruders into the space? If two males have adjoining territories, how do they react to each other? Is their reaction to each other different from that to nonneighbor dragonflies?

Territorial behavior consists of responses to the physical environment as well as to other animals. Determine some of the parameters of your fly's territory. What is the size and shape of the territory (including the vertical dimension)? Is there anything conspicuous about the boundaries or kinds of things included within the territory (vegetation, landmarks, etc.)? Does the male utilize a perch in his territory? Where is it located relative to territorial boundaries? How often does the male patrol his territory and how long does it take? Describe his behavior during patrolling. Attempt to observe several males, or pool data from the class, to determine patterns of territory shape, size, and topography, and patterns of territorial behavior.

Count the number of territorial males in the area you are observing and in any other suitable ones you can find. Is the density of territorial males the same in different areas? If not, what factors might cause the differences?

DISCUSSION

Consider the significance of territoriality to an individual and to the population to which it belongs. Defending a territory requires a great deal of time and energy, and, for a dragonfly, greatly increases the exposure of the territorial individual to predation. What are the advantages of possessing a territory? For many kinds of animals only a relatively fixed number of territories can be established in a suitable area; if a territory owner is removed, another animal quickly takes its place. With this type of social situation, what effects might territoriality have on the rate of reproduction and population size in a given area? In your further reading note the kinds of animals in which territoriality is found and the different kinds of territories maintained.

ADDITIONAL STUDIES

Observation of dragonflies may require a great deal of time, but if time is available several additional experiments may be carried out with them. Some possible studies are suggested here.

Introduction of Intruders. Capture several dragonflies at some other location. This is not a simple task, as dragonflies are rapid fliers and have good eyesight. Ambushing an individual on patrol or netting from behind are possible techniques. Be as gentle as possible with your captive as the wings are easily injured. Either of two methods may be used to study the effect of introducing a captive into an established territory. One is to release it within the territory and record the reaction of the territory owner. Another is to tie a loop of cotton thread around the thorax and tie about 1 to 1.5 meters of the thread to a long stick or dowel. The tethered dragonfly (either alive or dead, but preferably alive) may be dangled in the territory of the fly being observed. (The observer should be as still as possible.) Several pertinent questions can be answered by appropriate introductions. From what distance can the owner detect the intruder? Does the owner react differently to males, females, or dragonflies of other species? If the owner attacks the captive intruder, slowly move the captive away and determine how far away from the territorial boundary you can lead the owner. Does habituation to the intruder occur?

Removal of Territory Owners. Remove a male from its territory. How soon is the area taken over by a new dragonfly? If the original owner is released in the area will he regain the territory? Do territorial boundaries remain the same even though the owners change? If not, how do they differ in size, shape, boundary location, and so on?

Persistence of Territories. If you can observe a pond or a stream for a full day or return for several days, you can attempt to answer questions about the stability of the territories. How many territories are maintained around a body of water? Does the number change in the course of several days? For how long a time does a male maintain a territory? This will require marking the male in some way. One method is to capture him and put small distinguishing paint markings on his abdomen. Gently fold back both pairs of wings and hold the animal by them. Small pieces of grass, sedge, or sticks may be used to paint stripes 1 or 2 mm wide on the top of the animal's abdomen. Try not to cover the flexure point. By varying colors and the position of the colors a number of distinguishing marks can be made. Does the marked male return to the same or a different area on subsequent days following release?

REFERENCES

Bick, G. H., and J. C. Bick, 1965. Demography and behavior of the damselfly, *Argia apicalis* (Say) (Odonata: Coenagriidae). *Ecology.* 46:461–472.

Borror, D. J., and D. M. DeLong, 1954. *An introduction to the study of insects,* pp. 101–121. New York: Holt, Rinehart, and Winston.

Borror, D. J., and R. E. White, 1970. *A field guide to the insects,* pp. 68–75. Boston: Houghton Mifflin.

Campanella, P. J., and L. L. Wolf, 1974. Temporal leks as a mating system in a temperate zone dragonfly (Odonata: Anisoptera) I: *Plathemis lydia* (Drury). *Behaviour* (in press).

Corbet, P. S., 1962. *A biology of dragonflies.* London: Witherby.

Corbet, P. S., C. Longfield, and N. W. Moore, 1960. *Dragonflies.* London: Collins.

Jacobs, M. E., 1955. Studies on territorialism and sexual selection in dragonflies. *Ecology* 36: 566–586.

Johnson, C., 1962. A description of territorial behavior and a quantitative study of its function in males of *Hetaerina americana* (Fabricus) (Odonata: Agriidae). *Canad. Entomol.* 92: 178–190.

Johnson, C., 1964. The evolution of territoriality in the Odonata. *Evolution* 18:89–92.

Kaufman, J. H., 1971. Is territory definable? *In* A. H. Esser (Ed.), *Behavior and environment,* pp. 36–39. New York: Plenum Press.

Moore, N. W., 1952. On the so-called "territories" of dragonflies (Odonata: Anisoptera). *Behaviour* 4:85–100.

Moore, N. W., 1964. Intra- and interspecific competition among dragonflies (Odonata). *J. Anim. Ecol.* 34:49–71.

Pajunen, V. I., 1966. Aggressive behavior and territoriality in a population of *Calopteryx virgo* l. (Odon., Calopterygidae). *Ann. Zool. Fenn.* 3:201–214.

JULIA CHASE
Livingston College, Rutgers University

Interspecific Aggression in Birds

Many passerine (perching) birds exhibit two distinct phases of social behavior each year. In the fall and winter they are generally gregarious, foraging and roosting in flocks that may include one or more species. In late winter, endocrine changes lead to intraspecific aggression and pair bonding. Individual breeding pairs establish territories from which they exclude all conspecifics and occasionally individuals of other species (Murray, 1971). This social isolation lasts until the last young are fledged in late summer, at which time the birds revert to their gregarious habits.

There are many benefits to flocking behavior (e.g., aid in detection of predators, food finding, etc.). On the other hand, direct competition for food is also increased in the flock. This competition can be dealt with in several ways: (1) the species can forage in separate microhabitats, (2) individual birds can fight for the food, or (3) individuals and/or species may establish stable dominance relationships in which the feeding priorities are established and subordinate birds simply defer to dominants who wish to feed.

In mixed-species bird flocks, certain species may be dominant over other species. The purpose of this exercise is to examine these interspecific dominance relationships. A single-perch feeder will be used to elicit direct competition between species. This technique will allow the investigator to study: (1) the behavior patterns in feeding and aggressive behavior in several bird species, (2) the relative dominance among the species observed and the type of hierarchy established, and (3) the basis for interspecific dominance in passerine birds.

METHODS

Subjects and Materials

If you have access to a well-established bird feeder, observe the birds until you have become familiar with the species that normally feed there. Feeders with a single perch ("satellite" feeders, Fig. 36.1) are best for this study. If you are setting up a feeder yourself, choose an area with few other feeders, and begin baiting the feeder several weeks in advance of the planned study so that the birds can learn its location. The only materials needed are a clipboard, field guide (for species identification), pencil, and paper.

Procedure

This study should ideally be conducted in midwinter, when competition for food is high and intraspecific territorial and reproductive fighting are low. Choose an hour when the birds are normally active (mornings and evenings are best) and observe at that time for approximately an hour for each of 10 days within a two-week period. This will minimize the effect of diurnal and seasonal variations in aggressive and feeding behavior. Keep a record of the number of visits made by members of each species. Record all "supplanting attacks" (see below), noting the species of both the individual attacked and the attacker. Be sure to record the date, hour, and weather conditions. If severe weather seems to discourage some species from feeding, omit these days from your data.

When a bird flies toward a single-perch feeder at which another bird is already feeding, the newcomer may either turn away and fly off, or displace ("supplant") the bird already at the feeder. In this latter type of interaction, there occasionally is a chase or combat, but more often the previously feeding bird simply flies off. Behaviorally, these so-called "supplanting attacks" consist of the intruder swooping down on the feeder and, in the genus *Parus* (chickadees and tits), "the attacker appears to glide for the last foot or two of his attacking flight and gives a momentary wings-raised posture on landing" (Hinde, 1952). The details of such attacks, however, will be hard to observe; you will have to establish behavioral criteria to distinguish actual supplants from random departures.

With a single-perch feeder, a supplanting attack may be defined simply as an interaction which meets the following two criteria: (1) the intruding bird physically takes the place of the feeding bird on the perch, and (2) the feeding bird makes its departure when the intruder is within 3 meters of the feeder. This distance is chosen arbitrarily and may need to be modified under certain conditions (e.g., if your feeder has a hedge 2 meters away in which the birds normally sit before feeding).

Supplanting attacks are much more difficult to define if the feeder is an open tray. These have many possible perches and the feeding bird may simply move to the side rather than fly away when the intruder approaches. Supplanting attacks should be defined as rigorously as possible in this situation, using criteria such as the following: (1) dyadic interactions only (i.e., no more than two birds present, the bird feeding and the intruder), (2) the feeding bird actually leaves the feeder, and (3) this departure occurs within the 2 seconds before or the 5 seconds after the intruder lands. Again, you may have to modify these criteria to suit your needs.

Summarize your observations in Table 36.1 and line *a* of Table 36.2. Then calculate the relative frequency with which each species visited the feeder (line *b*, Table 36.2: the percentage of visits made by each species). From the first table, add the number of supplanting attacks made by each species (line *c*, Table 36.2) and calculate the percentage of the visits in which each species performed an attack (line *d*, Table 36.2).

Determine whether a given species tended to attack certain species more than others. If the attacks by species A were distributed randomly, the expected number of attacks by A on species B, for example, would equal the percentage of the visits to the feeder

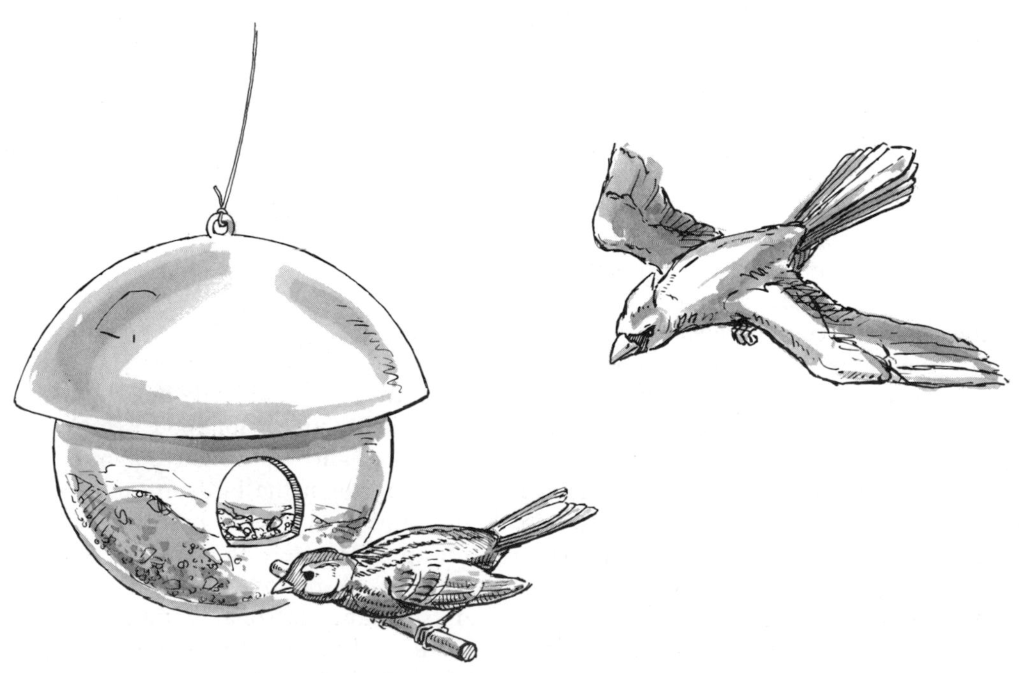

Figure 36-1. Supplanting attack at a satellite feeder.

Table 36-1. Sample data sheet for recording interspecies supplanting attacks.

Species attacked	*Attacking species*			
	A	*B*	*C*	*D*
A				
B				
C				
D				

made by species B (line *b* for the species attacked) multiplied by the total number of attacks that A made (line *c* for species A). In other words, the expected number of attacks on each species should be proportional to the amount of time that species spends on the feeder.

You may wish to use the chi-square test to determine the significance of the difference between the expected and observed number of attacks exhibited by various species.

QUESTIONS

1. Describe the agonistic behaviors observed. Was there any overt fighting? Did any species vocalize during these displacements? Which species visited the feeder most? Which performed the most supplanting attacks?

2. Are the majority of the supplanting attacks directed toward conspecifics or other species? Why do you think Hinde (1952) finds the percent of attacks on conspecifics to be higher in the spring than in winter?

3. Does any species consistently dominate another species? Do dominant species spend more time per visit on the perch?

4. Can these interspecific dominance relationships be arranged in a linear hierarchy (A>B>C>D) or are the relationships nonlinear (e.g., A>B>C but C>A)? What are some of the factors that may determine relative dominance between species?

5. What is the biological value of fighting? Does it surprise you that Hinde (1952) finds more fighting in closely integrated flocks than loosely bound ones? Why does he observe more fighting during winters of food scarcity than when food is abundant? What are the alternatives to fighting?

6. What are the ecological consequences of interspecific aggression? For example, what effect might it have on the foraging and nesting habits of a subordinate species (Morse, 1970)?

ADDITIONAL STUDIES

Students may wish to study whether sex, relative numbers of each species present at the feeder, and time of year affect the interspecific dominance relations.

Mixed-species flocks of birds can often be observed foraging in the field (e.g., woodland, beach, etc.). This presents an opportunity to study the general behavior of such flocks and interspecific dominance relationships in a natural setting. What is the size and composition of such groups? Do they have a leader? Are alarm calls heeded only by conspecifics or by all members of the flock?

The principles and methods outlined here can be used to study interspecific dominance hierarchies in small rodents, using an arena divided by a removable

Table 36-2. Sample table for recording relative frequencies of feeding bouts and attacks.

Category	*Species*				*Sum*
	A	*B*	*C*	*D*	
a) Number of visits to feeder					
b) Percent of visits made by each species					100%
c) Total number of attacks					
d) Percent of visits in which an attack is made					

partition (King, 1957; Miller, 1969; Baenninger, 1973), or even in different "species" of automobiles at a Y intersection (Dobb and Gross, 1968).

REFERENCES

Baenninger, L. P., 1973. Interspecific aggression in wild mice. *J. Comp. Physiol. Psychol.* 82:48–54.

Dobb, A. N., and A. E. Gross, 1968. Status of frustrator as an inhibitor of horn-honking responses. *J. Soc. Psychol.* 76:213–218.

Hinde, R. J., 1952. The behavior of the Great Tit (*Parus major*) and some other related species. *Behaviour* 2(Suppl.):1–199.

King, J. A., 1957. Intra- and inter-specific conflict of *Mus* and *Peromyscus. Ecology* 38:355–357.

Miller, W. C., 1969. Ecological and ethological isolating mechanisms between *Microtus pennsylvanicus* and *Microtus ochragaster* at Terre Haute Indiana. *Amer. Midland Natur.* 82:140–148.

Morse, D. H., 1970. Ecological aspects of some mixed species foraging flocks of birds. *Ecol. Monogr.* 40:119–168.

Murray, B. G., 1971. The ecological consequences of interspecific territorial behavior in birds. *Ecology* 52:414–423.

Sabine, W. S., 1949. Dominance in winter flocks of juncos and tree sparrows. *Physiol. Zool.* 22: 64–85.

Stokes, A. W., 1962. Agonistic behavior among blue tits at a winter feeding station. *Behaviour* 19:118–138.

ALAN M. BECK
Washington University

Behavior of Dogs: Canid Behavior in a Natural Setting

"Knowledge of the mere occurrence of wolves imparts to a district a wilderness charm otherwise lacking" (Stebler, 1944); a somewhat similar, but often overlooked, feeling of wild independence is sensed when watching a pack of dogs roam city streets at dawn. For the straying (owned) or stray (ownerless) dog, the city is the natural habitat, with man just another biotic component. We should be reminded that all animals are affected by man, or at least by man's pollution. Dogs are ideal for behavioral study for they are common, large, mostly diurnal, social, individually identifiable, tolerant of human proximity, and stay within relatively small home ranges.

Studies of the urban dog, a readily accessible and approachable animal, can provide training in behavior that then can be extended to other field situations and other canids.

Objectives could include cataloging the behavior of dogs, noting which behaviors are direct adaptations to the urban environment and which are typical of all canids, as well as comparing and contrasting dog–dog interactions with human–dog interactions.

In addition to its educational value, studies of dog behavior may be of immediate value as urban dog populations increase. There are many public health problems related to such populations, e.g., dog bite, diseases from dogs, garbage disruption, and fecal and urine contamination of the urban environment.

You should be familiar with some of the basics of canid behavior and communication. Beck (1971; 1973), Fox (1969; 1970; 1972a), Kleiman (1967), Schenkel (1967); Scott and Fuller (1965) are just a few of the readily available papers and books with descriptions and illustrations. Figure 37.1 is designed to give a general orientation to the social postures observed in dog–dog interactions.

METHODS

Subjects and Materials

Sections of the city should be surveyed for free-ranging dogs. Areas of high human density, low-rise housing, tenant rather than owner residency, and low income should be considered first. College campuses and other areas likely to support dog populations should be considered as well. Knowing the trash collection days is helpful (see Beck, 1973).

Equipment needed includes a field notebook, watch, and map of the area. A portable tape recorder, camera, and thermometer are useful.

Procedure

Students working in small groups (3 or 5 each) should meet about an hour after sunrise in the area selected. The early morning is recommended for it coincides with canine activity and human inactivity. Cities are safe and interesting before the rush hour.

A. Behavior of Individuals A specific individual could be selected from a group (or even a pet known to be let out every morning) and followed for several mornings or early evenings (see Beck, 1971). Traveling on foot is best but sometimes an automobile is useful.

A behavior profile should be generated for each animal, including when the animal is active (activity patterns), scent-marking patterns (urinating and defecating), the area traversed during normal activities (home range), foraging patterns, and behaviors associated with meeting other dogs and human beings (see Fig. 37.1). Frequency and duration of behaviors should be recorded and the points of occurrence located on a map of the area.

B. Behavior in a Localized Area Dog activity and social interactions can be observed by spending several hours each day at a location commonly frequented by dogs (e.g., an alley in which trash is normally found). The effects of various environmental components could be quantified, e.g., time, weather, trash collection, and season. Remember that other animals share the same environment. The behavior of various species of birds as well as that of cats, rats, squirrels, and man should be included. Do dogs respond to these animals, and vice versa? If evening observation is inconvenient, areas around trash could

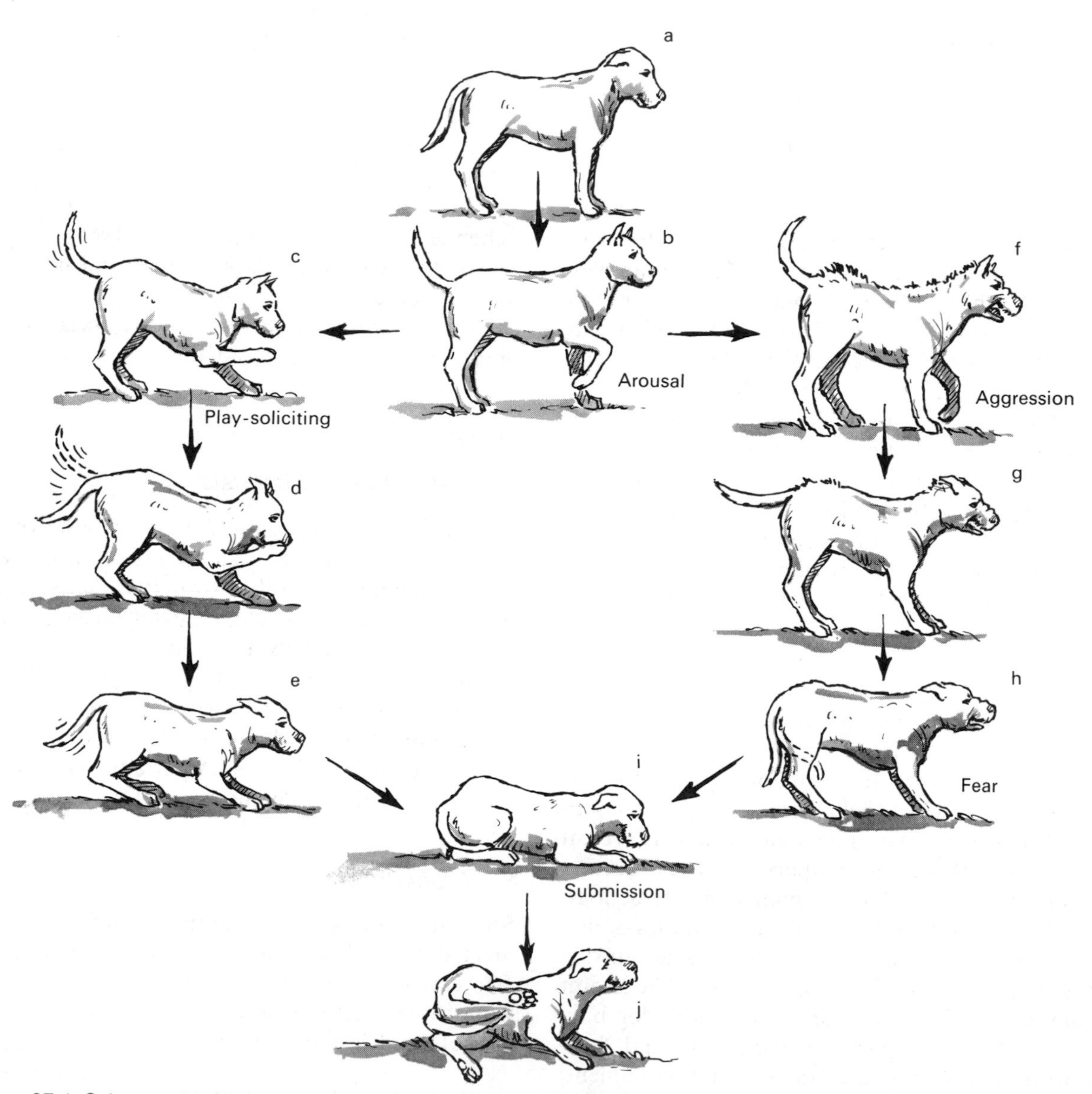

Figure 37-1. Schema of body language and expressive social responses of the dog. (*a–b*) Neutral to alert attentive positions. (*c*) Play-soliciting bow. (*d–e*) Active and passive submissive greeting; note tail wag, shift in ear position and of distribution of weight on fore and hind limbs. (*f-h*) Gradual shift from aggressive display to ambivalent fear-defensive-aggressive posture. (*i*) Passive submission with (*j*) rolling over and presentation of inguinal-genetal region. (After Fox, 1972a.)

be swept clean, then covered with sawdust. Return the next morning to check for dog and cat tracks and droppings.

C. Flight Distance Flight distance in dogs can be quantified by simply approaching animals. The distance at which a dog barks and flees can be measured with a tape measure or camera range finder. What effects do weather, pack size, number and behavior of observers have on flight distance?

A word of warning: there is always the danger of dog bite. It can be minimized by working in groups, avoiding private property, and being prepared to stand one's ground. Feigning to throw an object almost always works. In these ways, dog–human interactions during observation can actually be made safer than those that occur naturally.

QUESTIONS

1. What are the social and ecological parameters that constitute a good free-ranging dog "habitat"?
2. What is the role of scent-marking? Why do males leg-lift and why do dogs sometimes scratch backwards after urinating? (Beach and Gilmore, 1949; Bekoff, pers. comm.[1]; Berg, 1944; Hart, 1967).
3. What are the roles of dominance in the dog and at what times do they appear? Are there tests to determine dominance?
4. What factors lead to pack information? Which behavioral patterns are directly related to living in packs?

ADDITIONAL STUDIES

The local zoo may be used to observe the general repertoire of canid behaviors as well as to compare the behavior of different species (see Kleiman, 1967; Fox, 1973).

Newborn puppies may be available from pet owners or from the local animal shelter. Raising a dog and quantifying its behavioral patterns as they develop can be of great value in understanding the ontogeny of behavior (see Bekoff, 1972; Fox, 1972b; Scott and Fuller, 1965).

As an extended project, study campus dogs (Bekoff, pers. comm.[1]). It is easy to get to know individual animals. You may wish to study dominance relations in a group or breed differences in behavior. What are the sources of food, water, and shelter? Do dogs that are let out to run (straying) behave differently from ownerless strays?

[1]Bekoff, M. Scent marking by free-roaming dogs on the Washington University Campus, St. Louis, Mo. 1972–1974. Available on request from author.

REFERENCES

Beach, F. A., and R. W. Gilmore, 1949. Response of male dogs to urine from females in heat. *J. Mammology* 30:391–392.

Beck, A. M., 1971. The life and times of Shag, a feral dog in Baltimore. *Natur. Hist.* 80(8):58–65.

Beck, A. M., 1973. *The ecology of stray dogs: A study of free-ranging urban animals.* Baltimore: York Press.

Bekoff, M., 1972. The development of social interaction, play, and metacommunication in mammals: An ethological perspective. *Quart. Rev. Biol.* 47(4):412–434.

Berg, I. A., 1944. Development of behavior: The micturition pattern of the dog. *J. Exp. Psychol.* 34:343–368.

Fox, M. W., 1969. The anatomy of aggression and its ritualization in canidae: A developmental and comparative study. *Behaviour* 35:242–258.

Fox, M. W., 1970. A comparative study of the development of facial expressions in canids; wolf, coyote, and foxes. *Behaviour* 36:49–73.

Fox, M. W., 1972a. *Understanding your dog.* New York: Coward, McCann and Geoghegan.

Fox, M. W., 1972b. *Integrative development of brain and behavior in the dog.* Chicago: University of Chicago Press.

Fox, M. W., 1973. *Behavior of wolves, dogs and related canids.* New York: Harper and Row.

Hart, B. L., 1967. Sexual reflexes and mating behavior in the male dog. *J. Comp. Physiol. Psychol.* 64:388–399.

Kleiman, D., 1967. Some aspects of social behavior in the canidae. *Amer. Zool.* 7(2):365–372.

Schenkel, R., 1967. Submission: Its features and function in the wolf and dog. *Amer. Zool.* 7(2):318–329.

Scott, J. P. and J. L. Fuller, 1965. *Genetics and social behavior of the dog.* Chicago: University of Chicago Press.

Stebler, A. M., 1944. The status of the wolf in Michigan. *J. Mammalogy* 25(1):37–43.

DAVID P. BARASH
University of Washington

Human Ethology and the Concept of Personal Space

A variety of popular books (Lorenz, 1966; Adrey, 1970; Morris, 1967) have recently attempted to interpret human behavior in ethological terms. Although many of the ideas in these books are fascinating and controversial, they unfortunately do not readily lend themselves to scientific testing. However, the normative behavior of the animal *Homo sapiens* is just as amenable to objective analysis as is the behavior of nonhuman animals—if we choose the right behaviors to evaluate. Many animals demonstrate the existence of a characteristic "personal space," a three-dimensional volume surrounding each individual within which the approach of another elicits unease or retreat (Condor, 1949; Hediger, 1955). Human beings also maintain characteristic levels of personal space, the quantitative details varying with the nature of the social interaction and the cultural norms of the individuals concerned (Hall, 1969). In this exercise you will investigate spatial behavior in that aberrant primate, *Homo sapiens.*

METHODS

Subjects and Materials

This field study can be conducted almost anywhere. Your work will include both observation and simple experiments. The reading room of a university or municipal library is an ideal study site, and will be assumed to be the locale of the following investigation, although, with appropriate adjustments, restaurants, park benches, theaters, or buses may be substituted. A watch with a sweep second hand is the only equipment necessary.

Procedure

First, construct a map of the seating area, showing the placement of tables and the number of individuals at each table. Compare the results with that expected by random (e.g., Poisson) distribution (Stilson, 1966) to determine whether the spacing was greater or less than would be expected by chance alone. Alternatively, record the percentage of tables that must be occupied by at least one person before people begin to be found two (or three) at a table.

If the study area is crowded, concentrate your observations on the spacing pattern of each table. Choosing individuals at random, compare the frequency with which spaces adjacent to each are occupied with the frequency of occupants in nonadjacent spaces.

In order to investigate experimentally the effects of invasions of personal space upon subsequent behavior, control data must first be obtained. Select individuals alone at separate tables and, observing from a discrete distance, record the number remaining seated during 3-minute intervals for 30 minutes. Obtain data for at least ten individuals. Your experimental method consists of approaching other randomly selected solitary subjects, sitting at different predetermined distances from them, and recording the number remaining at 3-minute intervals. By comparing these results, you have a measure of the effect of your invasion of the subjects' personal space. (It is essential that each test be conducted with a different subject, that the subject be the sole occupant of his table and that the subject be unaware of his participation in the experiment.) The data may be tabulated as shown in Table 38.1.

Compare the effect of the following conditions of experimental approach: across table, diagonal; across table, facing; same side, adjacent seats empty; same side, adjacent. In addition to recording the time the subject remains seated, you might also note the frequency of chair movements and occurrence of verbal protests (faint-hearted experimenters please note: the latter response is surprisingly rare!). Compare your results with those reported in the literature (Felipe and Sommer, 1966; Barash, 1973).

QUESTIONS

1. What is the selective advantage to the maintenance of personal space among nonhuman animals? Human beings?

2. Distinguish between personal space and territorial behavior.

3. What relationship would you expect between personal space and individual dominance?

4. Would you expect the individuals to demonstrate the same degree of personal space under all circumstances? Under what conditions would you expect the maintenance of greater personal space? Less?

ADDITIONAL STUDIES

You can compare the responses of male *vs* female, student *vs* professor, black *vs* white, studious *vs* day-dreaming subject, and so on. If possible, interview departing subjects to determine possible correlations between personal-space manifestations and ethnic background, urban *vs* rural upbringing, and so forth (see Leibman, 1970).

The field of human ethology is brand new and "wide open." With a little thought and imagination, you should be able to identify many other human behavior patterns that might profitably be investigated in an objective, quantitative manner (see

Table 38-1. Data sheet for recording the effects of the invasion of personal space on the spatial behavior of humans.

Number of minutes following onset of observations	*Number of people remaining seated*	
	No approach	*Approach*
0		
3		
6		
9		
12		
15		
18		
21		
24		
27		
30		

Sommer and Becker, 1969; DeLong, 1970; Barash, 1972).

REFERENCES

Ardrey, R., 1970. *The social contract.* New York: Atheneum.

Barash, D., 1972. Human ethology: the snack-bar security syndrome. *Psychol. Rep.* 31:577–578.

Barash, D., 1973. Human ethology: personal space reiterated. *Environ. Behav.* 6:69–72.

Condor, P. J., 1949. Individual distance. *Ibis* 91: 649–655.

DeLong, A., 1970. Dominance-territorial relations in a small group. *Environ. Behav.* 2:170–191.

Felipe, N., and R. Sommer, 1966. Invasions of personal space. *Soc. Probl.* 14:206–214.

Hall, E., 1969. *The hidden dimension.* New York: Doubleday.

Hediger, H., 1955. *Psychology of animals in zoos and circuses.* London: Butterworth.

Leibman, M., 1970. The effects of sex and race norms on personal space. *Environ. Behav.* 2:208–246.

Lorenz, K., 1966. *On aggression.* New York: Harcourt, Brace and World.

Morris, D., 1967. *The naked ape.* New York: Jonathan Cape.

Sommer, R., and F. D. Becker, 1969. Territorial defense and good neighbor. *J. Pers. Soc. Psychol.* 11(2):85–92.

Stilson, D., 1966. *Probability and statistics in psychological research and theory.* San Francisco: Holden-Day.